FORAGE GROUPS

Edited by **Ricardo Loiola Edvan**
and **Edson Mauro Santos**

Forage Groups
http://dx.doi.org/10.5772/intechopen.75997
Edited by Ricardo Loiola Edvan and Edson Mauro Santos

Contributors

Steffen Beck-Broichsitter, Heiner Fleige, Rainer Horn, Horst Gerke, James R. Kiniry, Sumin Kim, Norman Meki, Mari-Vaughn Johnson, Sharon Clay, Heidi Reed, Michelle Ohrtman, Alexander Smart, David Clay, Albrecht Glatzle, Ricardo D. Améndola-Massiotti, Renato González-Ortiz, Luis Alberto Miranda-Romero, Juan A. Burgueño-Ferreira, Pedro Topete-Pelayo, Zivanayi Matore, Srinivasan Ramakrishnan, Tejveer Singh, Sanath Kumar Mahanta, Vikas Tyagi, A.K. Roy, Ignatius V. Nsahlai, Mehluli Moyo, Katamzi Masande, Siyabonga Bhiya

First published in London, United Kingdom, 2019 by IntechOpen
IntechOpen is the global imprint of INTECHOPEN LIMITED, registered in England and Wales, registration number: 11086078, The Shard, 25th floor, 32 London Bridge Street
London, SE19SG – United Kingdom
Printed in Croatia

British Library Cataloguing-in-Publication Data
A catalogue record for this book is available from the British Library

Additional hard copies can be obtained from orders@intechopen.com

Forage Groups, Edited by Ricardo Loiola Edvan and Edson Mauro Santos
p. cm.
Print ISBN 978-1-78985-681-1
Online ISBN 978-1-78985-682-8

For EU product safety concerns:
IN TECH d.o.o., Prolaz Marije Krucifikse Kozulić 3, 51000 Rijeka, Croatia,
info@intechopen.com or visit our website at intechopen.com.

Meet the editors

Dr. Ricardo Loiola Edvan received his Doctorate in Animal Science from the Ceara Federal University, 2013; his Masters degree in Animal Science from the Paraiba Federal University, 2010; and his Graduate degree in Agronomy from the Ceara Federal University, 2006. He is a Professor in Animal Science, Veterinary and Agronomy courses at the Piaui Federal University, with a major in Forage and Pastures areas. He is a member of the Animal Science Post-graduate Program in the Piaui Federal University and coordinator of research and technology development projects supported by private and public institutions.

Dr. Edson Mauro Santos received his Doctorate in Animal Science from the Vicosa Federal University, 2007; his Masters degree in Animal Science from the Rural Rio de Janeiro Federal University, 2002; and his Graduate degree in Animal Science from the Rural Rio de Janeiro Federal University, 2006. He is an Animal Science Professor at the Paraiba Federal University, with a major in Forage and Pastures areas. He is a member of the Animal Science Post-graduate Program in the Paraiba Federal University and coordinator of research and technology development projects supported by private and public institutions.

Contents

Preface

The book **Forage Groups** gathers information about forage plants that are used in different regions around the world as animal food. This book contains data from studies with native and cultivated species and presents information about cultivation and use of these forage plants, in a clear and direct way. The target readers are researchers, students, and farmers who will be able to apply the present knowledge on forage cultivation, harvesting, and conservation. Another interesting point of view is the use of legumes in animal feed, beyond bringing a crucial analysis about carbon dioxide – CO_2 produced in the pastures. The authors are from many countries around the world and this allows an analysis and exchange of information. The editors would like to thank the researchers and their staff and institutions for their time and knowledge.

Dr. Ricardo Loiola Edvan
Animal Science Department
Piaui Federal University
Teresina, Brazil

Dr. Edson Mauro Santos
Animal Science Department
Paraiba Federal University
Teresina, Brazil

Chapter 1

Effectiveness of Grassland Vegetation on a Temporary Capped Landfill Site

Steffen Beck-Broichsitter, Heiner Fleige,
Horst Gerke and Rainer Horn

Additional information is available at the end of the chapter

http://dx.doi.org/10.5772/intechopen.80324

Abstract

We studied the effectiveness of grassland vegetation of a temporary capping system consisting of differently compacted boulder marl and its impact on the water balance components. This study presents the modelled water balances for the period between 2008 and 2015, performed with HELP 3.95 D (German edition). The model requires landfill design and weather data as well as soil physical and evapotranspiration parameters including the leaf area indices and evaporative zone depth with regard to the grassland vegetation. The modelled average annual actual evapotranspiration rates ranged between 277 and 390 mm year-1 or rather 33 and 66% of the annual precipitation (10-year average of 728 mm). The actual evapotranspiration rates are strongly influenced by the maximum leaf area indices that increased between 2008 and 2015 from 1.0 to 3.5 as well as the evaporative zone depth that also increased from 20 cm in 2008 to 50 cm in 2015. The empirical-mathematical–based HELP model is a useful option to successfully determine the water balance components of a landfill capping system under the given weather and site conditions including the development of the grassland vegetation.

Keywords: HELP model, water balance, actual evapotranspiration, leachate generation, vegetation growth

1. Introduction

In a global perspective, landfill sites still represent the major option of waste disposal not only in developing countries [1]. In Germany, the qualitative criteria of landfills are legally fixed according to the [2] and define the vegetative and technical standards for engineered barriers [3].

In case of this study, semipermeable, temporary capping systems intend a specific shutdown of the bioreactor, containing heterogeneous wastes and different amounts of biodegradable material, through controlled infiltration of precipitation into the waste body [4] and also allow biogas extraction [5].

Temporary capping systems regularly consist of a recultivated layer, a drainage layer, and a sealing layer consisting of mineral substrates or in combination with polymers [6]. The major aim of the recultivated layer is to restrain landfill gas migration and to minimise leachate generation (precipitation contaminated with heavy metals or polycyclic hydrocarbons) by a high water storage capacity in combination with a distinct evapotranspiration rate from the vegetation and soil surface [3, 7].

Therefore, the choice of a locally adapted vegetation type (grassland, shrubs, forest) is essential to ensure high evapotranspiration rates (grassland: 450–550 mm $year^{-1}$), a quick vegetation establishment (erosion protection, slope stability), and avoid deep shrinkage-induced cracking (capillary rise from deeper horizons) and rooting to protect the sealing layer as last barrier above the waste body depending on the thickness of the recultivated layer [4, 8–10].

The functional requirements of the vegetation in the nutrient and water availability considering a proper air capacity and plant available water capacity [2], whereby the technical challenges in landfill construction, compacted installation versus loose installation of mineral substrates, can significantly influence the growth conditions of the vegetation [3].

The effectiveness of the vegetation can be assessed by the water balance or rather the leachate generation under the specific climate and soil conditions [4, 11, 12]. There are several modelling approaches of landfill capping systems, with and without polymers, combining water balance calculations with the predominant statistical-empirical Hydrologic Evaluation of Landfill Performance (HELP) model [13] or numerical models like Finite Element subsurface FLOW system (FEFLOW) [14]. Such predictive models can be used to support the planning of a landfill and/or to optimise the particular system from an economic point of view [12] and to verify the long-term hydraulic stability of a final capping system.

This study presents modelled water balance data and in particular the annual leachate rate of the Rastorf landfill during an 8-year period in the context of (a) grassland vegetation and (b) local weather conditions.

2. Materials and methods

2.1. Study site and weather conditions

The Rastorf landfill (lat. 54° 16′N, long. 10° 19′E) in Schleswig-Holstein (Northern Germany) was actively operated from February 1977 to May 2005 with a total area of 105,000 m^2 and about 2.0 million tons of municipal domestic wastes were deposited in it (**Figure 1**).

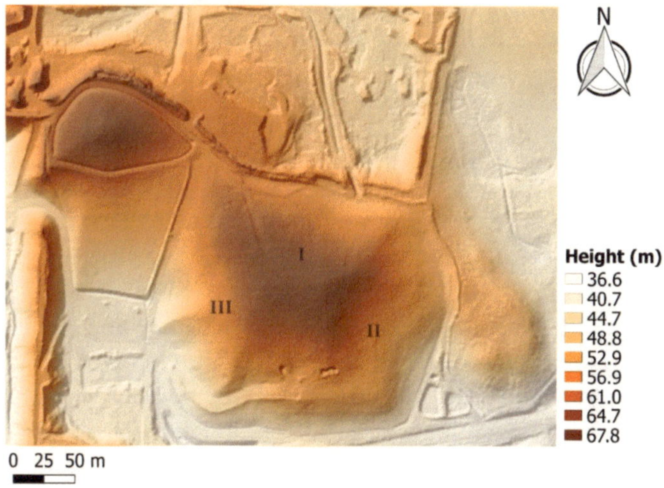

Figure 1. Digital elevation model of the Rastorf landfill with the temporary capped area (section I–III) [15].

The temporary capped area of nearly 75,000 m^2 with three sections (I: 21,275 m^2, II: 29,961 m^2, III: 22,208 m^2) consists of three mineral layers (boulder marl) with a partially permeable recultivated layer (humus topsoil: 40 cm, humus-poor subsoil: 30 cm) and, below this layer, is a low permeable, 30 cm thick mineral sealing layer, which serves as a water and root barrier to prevent leachate formation and the groundwater contamination. The bottom layer consists of hardly permeable up to 20 m thick clay. A high-density polymer of 2.5 mm thickness and a drainage system above the bottom layer collects the leachate before the treatment by inverse osmosis (**Figure 2**).

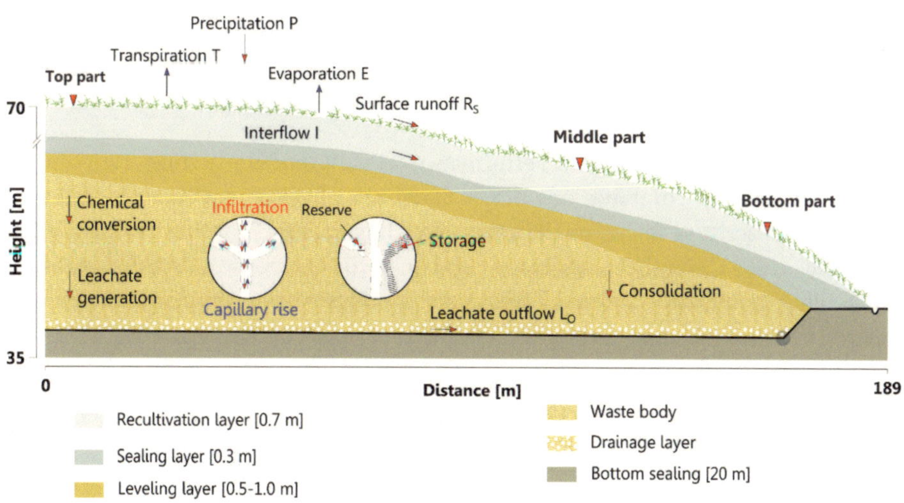

Figure 2. Schematic cross section through the temporary capped area with water balance components, data logger and measuring devices in 20, 50, 80 and 100 cm depth.

Area	I	II	III
Average slope gradient (°)	7 ± 3	14 ± 3	16 ± 4
Average slope length (m)	99 ± 65	48 ± 23	69 ± 4
Exposure	N/NE	SE	SW

The symbol ± corresponds to the standard deviation.

Table 1. Average slope gradient, slope length, and exposure of the sections I–III.

The maritime, semi-humid climate in Rastorf is characterised by an annual precipitation rate which is in the long-term average regularly between 6 and 9 months year-1 higher than the potential evapotranspiration rate [16]. The local weather conditions also affect water balances of the landfill capping system with 10-year average precipitation rates of 728 mm and, between 2012 and 2015, an average annual temperature of 9.0°C. The slope gradient varies between 7 and 16° and the slope length between 48 and 99 m (**Table 1**).

2.2. Laboratory measurements

In 2012, more than 160 undisturbed soil cores (100 cm^3) were sampled in the capping system in vertical (90°) and horizontal (0°) direction in area I (54°28′20″N, 10°32′60″E), II (54°28′11″N, 10°32′71″E) and III (54°28′08″N, 10°32′75″E) in depths of 0.2, 0.5 and 0.8 m. The saturated hydraulic conductivity (Ks) was measured under instationary conditions (n = 10 per depth) according to [17]. The pore size distribution (n = 7 per depth) was determined by a combined pressure plate (saturated, −6, −30) and − 1500 kPa ceramic vacuum outflow method as well as oven-dried at 105°C, respectively [18].

2.3. Hydraulic Evaluation of Landfill Performance (HELP) model

The Hydraulic Evaluation of Landfill Performance (HELP) model is a quasi two-dimensional hydrologic model which combines one dimensional soil physical and hydrological processes in (a) vertical direction and (b) lateral direction according to [13]. Thus, the model requires data of the landfill design, weather conditions, and material properties such as porosity, field capacity, wilting point and saturated hydraulic conductivity as input parameters [19]. In addition, the evaporative zone corresponds to the root depth of the vegetative cover and was calculated to quantify the maximum soil depth from which water can be removed through evapotranspiration [12].

With respect to the landfill design data, the upper part of the recultivation layer (0–0.4 m) was classified as vertical percolation layer, the bottom part (0.4–0.7 m) was conducted as lateral drainage layer to take into account the lateral saturated hydraulic conductivity. The sealing layer was classified as barrier soil liner.

The HELP model was validated with actual landfill data with respect to field and laboratory measurements according to [20].

2.4. Estimation of the water balance components of the Rastorf landfill

The HELP model was validated with actual landfill data with respect to field and laboratory measurements according to [20]. The leachate rate (L) was calculated as follows:

$$L(t_i) = P(t_i) - ET(t_i) - R(t_i) - D(t_i) \pm \Delta S(t_i) \tag{1}$$

where: L = leachate rate, P = precipitation, ET = actual evapotranspiration (including interception), R = runoff, D = lateral drainage (interflow) and ΔS = change in soil moisture content in mm year-1 and m3 and it is the time step, performed from January 1, 2012 until December 31, 2015.

A separate water balance was modelled for each area (I–III) and a weather station located close to the landfill recorded the actual meteorological data such as precipitation (uncorrected), air temperature, wind speed, wind direction, air pressure, air moisture, and relative humidity on daily basis. The global solar radiation was calculated on the basis of [21].

In addition, the wind speed was measured in 10 m height and a logarithmic approximation was used to calculate the wind speed for 2 m height. The leaf area index (LAI) was calculated on the basis of the quarterly measured average vegetation height (h) in 8–10 repetitive transects (1 m^2) per area with a folding ruler according to [21]:

$$LAI = 24 \cdot h \tag{2}$$

The average root intensity was determined annually on the basis of repetitive soil profile images in the three areas with the colour threshold method using ImageJ software [22] and classified according to [23].

2.5. HELP modules

The water balance calculations based on analytical and empirical equations, while a detailed description is shown in [19, 24]. With regard to the atmospheric boundary conditions, the method used in the HELP 3.95 D for calculating evapotranspiration was designed according to [25].

The potential evapotranspiration consists of (a) evaporation of surface water (primarily evaporation of intercepted water, besides this evaporation of snow), (b) soil evaporation, and (c) plant transpiration computed by a simplified approach of [26]:

$$E_{o_i} = \frac{PENR_i + PENA_i}{L_v} \tag{3}$$

$$L_v = \begin{cases} 59.7 - 0.0564T_{c_i} & \text{for water} \\ 67.67 - 0.0564T_s & \text{for snow} \end{cases} \tag{4}$$

where: E_{oi} = potential evapotranspiration on day i (mm), PENRi = radiative component of the Penman equation on day i (langleys), PENAi = aerodynamic component of the Penman equation on day i (langleys), L_v = latent heat for vaporisation (for evaporating water) or latent heat of fusion (for evaporating snow) in langleys per mm and T_s = snow temperature (°C).

The actual evapotranspiration (ETa) was mainly calculated by an approach of [25] using a model of vegetation growth and decay by [27]. Thus, the vegetative growth and decay sub-model included in HELP was taken from the model SWRRB [27]. The ETa is limited by the water availability at the landfill surface and the maximum depth of the evaporative zone according to [20]. Therefore, the plant available water capacity inside the evaporative zone (field capacity–wilting point) can only be removed by evapotranspiration, while the field capacity (US: −330 hPa) is the lowest soil water content to allow unsaturated vertical flow (drainage) within the evaporative zone [28]. The capacity of the interception storage and the interception height were calculated following Hoyningen-Huene (1983), modified and adapted to German standards by [28].

The area factor v was implemented in the modelling approach and corresponds to the ratio of the monthly sums of the global solar radiation (R_s) on inclined and horizontal reception areas consider the exposure and the inclination angle (°) and a corrected albedo of 0.23 in the summer-half (05/01-10/31) and in the winter-half (11/01-04/30) under climatic conditions in Germany [29].

The vertical percolation (drainage) is estimated using the equation for the unsaturated hydraulic conductivity (Eq. 4) which is based on [30]. The saturated lateral drainage is modelled by a steady-state solution of the Boussinesq equation in combination with the Dupuit-Forchheimer (Forchheimer, 1930) assumptions, which take into account the Ks value of the drainage layer. The unsaturated conductivity for each soil layer was calculated as follows:

$$K_u = K_s \left[\frac{\theta - \theta_r}{\Phi - \theta_r}\right]^{3+\left(\frac{2}{\lambda}\right)} \tag{5}$$

where: K_u = unsaturated hydraulic conductivity (cm s^{-1}), K_s = saturated hydraulic conductivity (cm s^{-1}), θ = actual volumetric water content (m^3 m^{-3}), θ_r = residual volumetric water content (m^3 m^{-3}), Φ = total porosity (m^3 m^{-3}) and λ = pore-size distribution index (−).

Therefore, θ_r is the amount of water remaining in a layer under infinite capillary suction and was estimated as follows [24]:

$$\theta_r = \begin{cases} 0.6\,\text{WP} & WP < 0.04 \\ 0.014 + 0.25\,\text{WP} & WP \geq 0.04 \end{cases} \tag{6}$$

where: WP = volumetric wilting point (m^3 m^{-3}).

The leakage rate depends upon the depth of the water-saturated soil (head) above the base of the layer, the liner thickness and the Ks value of the barrier soil. So, the leakage occurs whenever the moisture content of the layer above the liner is greater than the field capacity of the layer [19, 24].

In addition, the rainfall-runoff process is modelled using the SCS curve-number method with values above 0 up to 100, as presented in Section 4 of the National Engineering Handbook [31].

The curve numbers for the areas I–III were obtained under the terms of the surface slope, the slope length, and the vegetation cover and also modified according to the previous sensitivity analysis. The SCS-CN method based on the following basic form [32]:

$$R = \begin{cases} \frac{(P - I_a)^2}{P - I_a + S} & P > I_a \\ 0 & P \leq I_a \end{cases} \tag{7}$$

where: R = runoff (m^3), P = precipitation (m^3), S = potential maximum soil moisture retention auf the runoff begins (m^3) and I_a = initial abstractions (sum of interception + evapotranspiration + infiltration + depression storage) in m^3. The retention parameter S is transformed into a curve number (CN) with following relationship [24]:

$$CN = \frac{1000}{S + 10} \tag{8}$$

The lateral drainage layer required information about the maximum drainage length as length of the horizontal projection of a representative flow path and the drain slope for the areas I–III [30]. The lateral drainage equation can be described as follows [19]:

$$y^* = \frac{d^2y^*}{dx^{*2}} + \left(\frac{dy^*}{dx^*}\right)^2 + (\tan\alpha)\frac{dy^*}{dx^*} = \frac{q_D^*}{\cos^2\alpha} \tag{9}$$

where: $x^* = x/L$ (nondimensional horizontal distance), $y^* = y/L$ (nondimensional depth of saturation above liner), $q_D^* = q_D/K_D$ (nondimensional lateral drainage rate) with K_D = saturated hydraulic conductivity of the drain layer (cm/s) and α = inclination angle of the liner surface.

2.6. Model calibration and sensitivity analysis

The validity of the data used as input and output values for the comparison of observed and modelled data is of major importance [20]. Therefore, the sensitivity analysis, calibration, and validation for the period from 2008 to 2015 were performed in a previous study on the basis of input and output values of the HELP model [28].

Therefore, an increasing evaporative zone depth from 10 to 100 cm can increase the actual evapotranspiration up to 100 mm $year^{-1}$; an increasing LAI from 1 to 5 can increase the ETa values up to 85 mm $year^{-1}$. Additionally, an increasing slope of the drainage layer from 2–30% can reduce the annual leachate rate of about 25%.

The associated calibration study made it necessary to implement a lateral drainage layer instead of a vertical percolation layer in 0.4–0.7 m depth to take into account the basic concept of the landfill capping system due to anisotropic Ks values of the compacted layer (see Section 2.4).

The correlation coefficient (r^2) is an index of goodness of fit between the observed and modelled data according to [33].

3. Results

3.1. Vegetation growth of the Rastorf landfill

The recultivated layer of the temporary capped area is used as pasture with a grass and clover mixture of flat-rooted, densely growing, and perennial grasses. The seed mixture used in 2008/2009 was composed as follows: 20% perennial ryegrass (*Lolium perenne*), 20% cocksfoot (*Dactylis glomerata*), 21% red fescue (*Festuca rubra*), 21% sheep fescue (*Festuca ovina*), 10% meadow grass (*Poa pratensis*), 8% white clover (*Trifolium repens*), and a biannual mowing is carried out. Nowadays, the total coverage of the grass and clover mixture varies between 85 and 100% across the landfill surface (**Figure 3**).

The species composition is significantly different from the initial seed mixture after several years of growth: 70–80% cocksfoot (*Dactylis glomerata*), red and sheep fescue (*Festuca rubra, ovina*), and meadow grass (*Poa pratensis*), respectively (**Figure 3**).

The white clover (*Trifolium repens*) was characterised by an area fraction of about 5% and perennial ryegrass (*Lolium perenne*) with an area fraction of about 10%, predominantly on the areas (1000 m^2) subsequently secured in 2013 because of the reduced vegetation growth with locally available compost made out of tree and shrub cutting (**Figure 4**).

Figure 3. Vegetation growth of the Rastorf landfill between 2008 (left) and 2015 (right).

Figure 4. Vegetation growth before (left) and after (right) compost application June 2013.

The landscape-ecological inventories and pedological excavations during 2013 and 2015 resulted in fine roots that were able to reach a maximum depth of 25–30 cm (flat rooting) and a weak to medium intensity (< 10 roots dm^{-2}), mainly along smaller hair or shrinkage cracks in the upper part of the recultivated layer (**Figure 4**). The subsequently secured areas showed deeper and pronounced rooting with depths of 35–40 cm (medium rooting) and a medium to strong intensity (11–20 roots dm^{-2}).

3.2. Weather data, vegetative period and leaf are index

The evapotranspiration parameters for the HELP model are summarised in **Table 2**. The average annual wind speed varied between 4.35 m s^{-1} in 2015 and 4.91 m s^{-1} in 2013 and the average relative humidity (%) between 70.6 and 87.3% in the spring and summer months and between 82.5 and 95.2% in the autumn and winter months (**Table 2**). Additionally, the maximum leaf area indices with values between 1.0 and 3.5 were chosen on the basis of average annual LAI measurements in March, May, and July and October, respectively.

3.3. Landfill design and soil physical parameters

The porosities of the boulder marl differ between 0.292 and 0.307 m^3 m^{-3} in the barrier soil layer and 0.317 and 0.356 m^3 m^{-3} in the drainage layer as well as the percolation layer. The FC values range between 0.175 and 0.213 m^3 m^{-3}, while the WP values varied between 0.117 and 0.167 m^3 m^{-3} (**Table 3**). The highest Ks values were identified in the drainage layer between

Year	2008	2009	2010	2011	2012	2013	2014	2015
Average annual wind speed (m s^{-1})	4.76	4.56	4.73	4.75	4.78	4.91	4.58	4.35
Evaporative zone depth (cm)	20	20	30	50	50	50	50	50
Maximum leaf area index (−)	1.0	2.0	2.0	3.5	3.5	3.5	3.5	3.5
Vegetative period (days)	262	345	219	231	230	220	266	255
Average relative humidity (%)								
1. Quarter	82.5	88.2	87.6	87.7	88.5	89.7	89.2	90.8
2. Quarter	70.6	71.2	77.5	73.8	78.1	79.7	80.8	79.0
3. Quarter	81.0	76.3	80.5	87.3	82.3	81.6	82.0	82.6
4. Quarter	91.1	89.4	93.1	93.9	94.6	92.8	95.2	93.5

Table 2. Input data for the HELP model: Evapotranspiration parameters (latitude 54.2°).

Study area and layer		Porosity	FC*	WP**	Ks	WC***	Slope length and gradient
		(m^3 m^{-3})	(m^3 m^{-3})	(m^3 m^{-3})	(m s^{-1})	(m^3 m^{-3})	(m)/(%)
I	Percolation layer	0.356	0.184	0.127	4.5E-06	0.212	62/12
	Drainage layer	0.317	0.206	0.136	5.6E-04	0.244	
	Barrier soil layer	0.292	0.175	0.121	3.7E-09	0.292	
II	Percolation layer	0.352	0.191	0.117	5.8E-06	0.259	44/28
	Drainage layer	0.327	0.213	0.147	6.3E-04	0.226	
	Barrier soil layer	0.302	0.196	0.143	6.1E-09	0.302	
III	Percolation layer	0.332	0.207	0.167	5.9E-06	0.215	52/30
	Drainage layer	0.325	0.196	0.139	5.8E-04	0.217	
	Barrier soil layer	0.307	0.213	0.149	3.6E-09	0.307	

Data of the three subcatchment areas (I–III), n = 7–10 undisturbed soil cores per layer for the average values of porosity, field capacity (FC), wilting point (WP) and saturated hydraulic conductivity (Ks), including initial water content (WC) and slope length and gradient.
*Field capacity (−33 kPa), **Wilting point, ***Water content at the beginning of the year 2012.

Table 3. Input data for the HELP model: landfill design and soil physical parameters.

5.6e-04 and 6.3e-04 m s^{-1}, lower values were determined in the percolation layer between 4.5e-06 and 5.9e-06 m s^{-1} and the barrier soil layer had values ≤6.1e-09 m s^{-1}.

3.4. Modelled water balance components of the Rastorf landfill between 2008 and 2015

In the study period between 2008 and 2015, the climatic water balance was positive (precipitation > evapotranspiration) and with regard to the German weather conditions, the outflow (2008–2010) and the actual evapotranspiration (2011–2015) were the greatest output values of the water balance (**Table 4**). The years 2012 and 2013 showed lower annual

Water balance [mm year^{-1}]	**2008**	**2009**	**2010**	**2011**	**2012**	**2013**	**2014**	**2015**
Precipitation	757	726	852	760	655	669	753	767
Potential evapotranspiration*	602	619	555	557	526	556	571	534
Actual evapotranspiration*	277	284	280	383	390	332	362	364
Outflow**	351	297	457	262	179	270	285	300
Δ soil moisture content	0	0.5	−0.5	1.5	0.2	−0.2	−0.3	−0.3
Leachate	149	137	116	103	84	70	109	105

*Including interception.
**Surface runoff and lateral drainage.

Table 4. Average annual water balance components for the period between 2008 and 2015.

precipitation rates with 655 and 669 mm, respectively, compared to the average annual precipitation rate of 728 mm. The winters of 2008–2015 were mostly mild and only had some snow.

The modelled average annual ETa values ranged between 33% in 2010 and 60% in 2012, and the outflow rates between 39% in 2009 and 54% in 2010 of the annual precipitation. The changes in soil moisture content with −0.3 and 1.5 mm year^{-1} were moderate and the modelled leachate rates ranged between 14 and 18% in 2008–2010, and between 11 and 15% in 2011–2015 of the annual precipitation (**Table 4**).

These drier phases are characterised by higher discrepancies between ETp and ETa up to 4.9 mm d^{-1}, especially in the warmer months between June and September (**Figure 5**). On the other side, the early warming phase during March to May showed moderate discrepancies of 0.6–2.7 mm d^{-1}, and the period October to February of the following year indicated mostly no discrepancies between the potential and actual evapotranspiration.

The ETa values ranged between 46 and 50%, and since 2011 between 60 and 69% of the ETp with the increasing depth of the evaporative zone (20, 30–50 cm) and the maximum leaf area

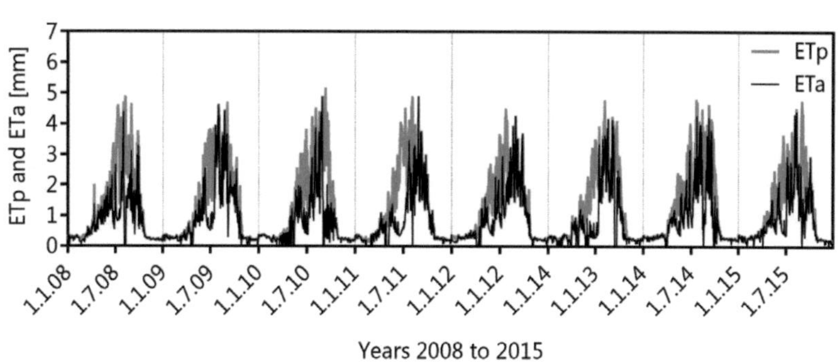

Figure 5. Average modelled potential and actual evapotranspiration rates (ETp, ETa) between 2008 and 2015 for the areas I–III.

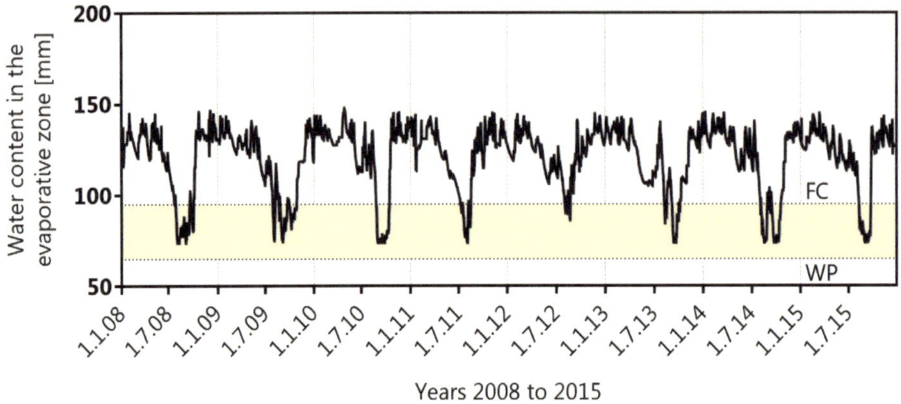

Figure 6. Average water content in the evaporative zone between 2008 and 2015 for the areas I–III. The dashed lines indicate the area between the HELP correlated field capacity (FC) and wilting point (WP).

indices (1, 2–3.5), respectively (**Figure 5**). The maximum depth complied with the part of the recultivation layer, in which the water content fluctuated intensely during the study period. The modelled water content in this evaporative zone appeared mostly above the field capacity of 95 mm, while longer phases during the vegetative period (July–September) nearly reached the wilting point of 65 mm, resulting in a decreased ETa capacity (**Figure 6**).

4. Discussion

The validity of the modelling results depends on the quality of the input data and related measurement methods that exhibit random errors depending on the site and weather conditions.

In this study, a weather station located close to the landfill recorded precipitation with a German Hellmann type gauge including wind-induced precipitation losses with an average underestimation of the actual annual precipitation of 10% [19]. Additionally, snow or rather snowmelt were no water balance factors during the study period between 2008 and 2015 under the weather conditions in Northern Germany.

The average annual actual evapotranspiration in Central Europe with an uncorrected precipitation rate of 700–800 mm (i.e., Rastorf landfill) is valued of approx. 450–550 mm for grassland vegetation with a good stand [34]. Therefore, the modelled annual ETa values, ranging between 277 and 390 mm are significantly lower than the mentioned ETa values for grassland vegetation. The modelled average annual ETa values ranged between 33 and 60% of the annual precipitation, but smaller than the ETa values of approx. Two-third of the annual precipitation in Hamburg (landfill Georgswerder) under approx. Comparable weather conditions as mentioned in [13]. The differences can be explained by the maximum leaf area index which strongly influences the evapotranspiration rate [3], while the HELP model assumed a constant LAI of 1, 2 or 3.5 for the whole year, respectively. On the other side, the daily average values of the wind speed do not reflect the actual wind conditions of an entire day [21] and the evaporative capacity of the wind-exposed Rastorf landfill must also be regarded as underestimated.

Longer phases during the vegetative period (July–September) nearly reached the wilting point of 65 mm, so, the evaporative zone (0.5 m) dried out more strongly and the transpiration capacity and thereby the ETa values of the grassland were restricted by (a) the inadequate water availability in the evaporative zone and (b) the limited water storage capacity, and (c) the limited capillary rise from deeper soil layers due to the compacted construction of the temporary capping system [5, 6]. Thus, phases with water contents below the critical field capacity of 95 mm should be as short as possible to prevent desiccation in the deeper layer, thus, the modelled water content is a first indicator to describe the hydraulic stability of the capping system.

Tree species or shrub vegetation (i.e., *Salix caprea* and *Ligustrum vulgare*) have a higher transpiration potential with ETa values of 600–700 mm $year^{-1}$ and are more effective than grassland to minimise the leachate generation of landfill capping systems [13]. However, more deep-rooted plants require thicker recultivation layers (2.0–3.0 m) to prevent shrinkage-induced crack formation in soil barriers due to desiccation and plant root penetration [9, 10]. Thus, the conflict of interest with regard to the choice of vegetation mainly depends on the local weather conditions, where robust grassland species should be preferred for locations with low precipitation [34], while more transpiring plant species are useful in more humid locations.

The modelled leachate rates were at a consistent level of 11–18% of the annual precipitation rate without significant deviations but exceeded the requirements of [2] at most 60 mm $year^{-1}$ 5 years after construction at the latest.

Otherwise, the modelled leachate rates indicate a sufficient percolation of water into the waste body to support the microbial processes [4]; between 2008 and 2017, the settlements of the waste body decreased from >20 to <4 cm $year^{-1}$, so, the temporary system fulfils its purpose.

The slightly varying annual leachate rates indicate the functionality of the temporary capping system; continuously rising leachate rates would be an indicator for shrinkage crack formation or root penetration in the sealing layer [10], thus, the capping system would be ineffective. So, the hydraulic stability of the temporary capping system and especially the barrier soil layer can be assumed as ensured.

In summary, the water balance model is not suitable to estimate more specific soil physical problems (i.e., preferential flow through cracks or root holes) of recultivation or sealing layers [5]. For an approved process description due to the model limitations, the numerical-based FEFLOW could be a more precise two-dimensional process description of the water fluxes of the Rastorf landfill in the saturated and unsaturated soil [14].

5. Conclusion

The HELP model is one of the most commonly used statistical-empirical approaches and is an useful option to successfully determine the leachate quantity of landfill capping systems and to proof which final capping system could be installed under the given weather and site conditions due to the statutory requirements.

The grassland vegetation of the Rastorf landfill changed in its plant-specific composition but is still effective to ensure moderate to high evapotranspiration rates and slope stability, while avoiding shrinkage-induced cracking and deeper rooting to protect the barrier soil or rather sealing layer. The future development depends on the intensity of wetting/drying cycles and further studies are required to improve the understanding of the long-term hydraulic stability of a mineral temporary capping system under climate change.

In order to finally proof the detailed water fluxes in structured landfill capping systems the more physically-based models will give more detailed insights into the variations in the soil water characteristics and the risk of crack formation in soil barriers due to desiccation and plant root penetration that may influence the functionality of it.

Acknowledgements

This research was supported by the Innovation Foundation Schleswig-Holstein and the ZMD Rastorf GmbH, Germany.

Conflict of interest

The authors do not declare conflict of interest.

Author details

Steffen Beck-Broichsitter[1]*, Heiner Fleige[2], Horst Gerke[1] and Rainer Horn[2]

*Address all correspondence to: steffen.beck-broichsitter@zalf.de

1 Research Area 1 "Landscape Functioning", Working Group "Hydropedology", Leibniz-Centre for Agricultural Landscape Research (ZALF), Eberswalder Straße, Müncheberg, Germany

2 Institute of Plant Nutrition and Soil Science, Christian Albrechts University Kiel, Germany

References

[1] Hoornweg D, Bhada-Tata P. What a waste: A global review of solid waste management. In: Urban Development Series, Knowledge Papers (15). World Bank, Washington, DC; 2012

[2] DepV – Decree on landfills (Landfill Ordinance): Decree on landfill sites and long-term storage facilities in the form of the resolution of the Federal Cabinet dated April 27th 2009

[3] Hauser VL. Evapotranspiration Covers for Landfills and Waste Sites. Boca Raton, FL: CRC Press: Taylor & Francis; 2008

[4] Rowe RK. Systems engineering: The design and operation of municipal solid waste landfills to minimize contamination of groundwater. Geosynthetics International. 2011;**18**(6):391-404

[5] Widomski MK, Beck-Broichsitter S, Zink A, Fleige H, Horn R. Numerical modelling of water balance for temporary landfill cover in North Germany. Journal of Plant Nutrition and Soil Science. 2015;**178**(3):401-412

[6] Beck-Broichsitter S, Fleige H, Horn R. Compost quality and its function as a soil conditioner of recultivation layers – A critical review. International Agrophysics. 2018;**32**(1):11-18

[7] Arnold S, Schneider A, Doley D, Baumgartl T. The limited impact of vegetation on the water balance of mine waste cover systems in semiarid Australia. Ecohydrology. 2014;**8**(3): 355-367

[8] Horn R, Baumgartl T. Dynamic properties of soils. In: Warrick AW, editor. Hrsg Soil Physics Companion. Boca Raton, Florida: CRC Press LLC; 2002. pp. 17-48

[9] Horn R, Stępniewski W. Modification of mineral liner to improve its long-term stability. International Agrophysics. 2004;**18**:317-323

[10] Witt KJ, Zeh RM. Crack due desiccation in cover lining systems. Phenomena and design strategy. International Workshop Hydro-Physico-Mechanics of Landfills, Grenoble 21–22.03. 2005; 2005

[11] Beck-Broichsitter S, Fleige H, Horn R. Waste capping systems processes and consequences for the longterm impermeability. In: Levin M, Kim HJ, Morel JL, Burghardt W, Charzynski P, Shaw RK, editors. Soils within Cities-Catena Soil Sciences. Stuttgart, Germany; 2017. pp. 148-152

[12] Pantini S, James Law H, Verginelli I, Lombardi F. Predicting and Comparing Infiltration Rates through Various Landfill Cap Systems using Water-balance Models – A Case Study. In: Proceedings ISWA World Congress 2013. Dettagli; 2013

[13] Berger K. On the current state of the hydrologic evaluation of landfill performance (HELP) model. Waste Management. 2015;**38**:201-209

[14] Diersch HJG. FEFLOW, Finite Element Subsurface Flow and Transport Simulation System – User's Manual/Reference Manual/White Papers. Berlin, Germany: WASY Ltd, Release 5.0; 2002

[15] QGIS Development Team. QGIS Geographic Information System. Open Source Geospatial Foundation; 2009

[16] Peel MC, Finlayson BL, McMahon TA. Updated world map of the Köppen-Geiger climate classification. Hydrology and Earth System Sciences. 2007;**11**:1633-1644

[17] Hartge KH. Ein Haubenpermeameter zum schnellen Durchmessen zahlreicher Stechzylinderproben. Zeitschrift für Kulturtechnik und Flurbereinigung. 1966;**7**:155-163

[18] Hartge/Horn. In: Horton R, Horn R, Bachmann J, Peth S, editors. Essential Soil Physics, an Introduction to Soil Processes, Structure, and Mechanics. Stuttgart, Germany: Schweizerbart Science Publishers; 2016

[19] Berger K, Schroeder PR. The hydraulic evaluation of landfill performance model. Version HELP 3.95 D. Software and Electronic Documents as PDF: User's Guide. Supplement to the Engineering Documentation of HELP 3.07. Germany: Institute of Soil Science, University of Hamburg; 2013

[20] Berger K. Validation of the hydrological evaluation of landfill performance (HELP) model for simulating the water balance of cover systems. Environmental Geology. 2000;**39**(11): 1261-1274

[21] Allen GA, Pereira LS, Raes D, Smith M. Crop evapotranspiration-guidelines for computing crop water requirements. In: FAO Irrigation and Drainage Paper 56. Rome, Italy; 1998. pp. 78-86

[22] Schneider CA, Rasband WS, Eliceiri KW. NIH image to image J: 25 years of image analysis. Nature Methods. 2012;**9**(7):671-675

[23] Boden Ad-Hoc-AG. Bodenkundliche Kartieranleitung. 5. Auflage. Hannover, Germany: Bundesanstalt für Geowissenschaften und Rohstoffe in Zusammenarbeit mit den Staatlichen Geologischen Diensten; 2005

[24] Schroeder PR, Dozier TS, Zappi PA, McEnroe BM, Sjostrom JW, Peyton RL. The hydrologic evaluation of landfill performance (HELP) model: Engineering Documentation for Version 3. EPA/600/R-94/168a. Risk Reduction Engineering Laboratory. Cincinnati, Ohio, USA: US Environmental Protection Agency; 1994

[25] Ritchie JT. A model for predicting evaporation from a row crop with incomplete cover. Water Resources Research. 1972;**8**(5):1204-1213

[26] Penman HL. Vegetation and Hydrology, Technical Comment No. 53, Commonwealth Bureau of Soils. Harpenden, England; 1963

[27] Arnold JG, Williams JR, Nicks AD, Sammons NB. SWRRB, a Basin Scale Simulation Model for Soil and Water Resources Management. College Station: Texas A&M University Press; 1990

[28] Berger K. Potential and limitations of applying HELP model for surface covers. Practice Periodical of Hazardous, Toxic, and Radioactive Waste Management. ASCE. July 2002;**6**(3): 192-203

[29] Unger H, Skiba M. Solare Strahlung auf geneigte Flächen. Sonnenenergie. 1998;**1**(98):48-50. (in German)

[30] Brooks RH, Corey AT. Hydraulic properties of porous media. In: Hydrol. Paper 3. Fort Collins, CO, USA: Colorado State Univ.; 1964

[31] USDA, Soil Conservation Service. National Engineering Handbook, Section 4, 115 Hydrology. Washington, D.C.: US Government Printing Office; 1985

[32] Soulis KX, Valiantzas JD. SCS-CN parameter determination using rainfall-runoff data in heterogeneous watersheds - the two-CN system approach. Hydrology and Earth System Sciences. 2012;**16**:1001-1015

[33] R Development Core Team. R: A Language and Environment for Statistical Computing. Vienna, Austria: R Foundation for Statistical Computing; 2008

[34] GDA-Empfehlung E2–32. Gestaltung des Bewuchses auf Deponien. Published in January 2010. http://www.gdaonline.de [Accessed: February 22, 2018]. (in German)

Chapter 2

Mob Grazing Results in High Forage Utilization and Reduced Western Snowberry Size

Heidi Reed, Sharon Clay, Alexander Smart,
David Clay and Michelle Ohrtman

Additional information is available at the end of the chapter

http://dx.doi.org/10.5772/intechopen.83402

Abstract

Mob-grazing strives to maximize forage utilization and minimize selective grazing by using high stocking densities in small paddocks for short durations (12–24 hr). Rotational-grazing uses low stocking densities for a longer time period, retaining about half of the original available forage; although selective grazing can occur. Three cattle (*Bos taurus* × *Bos indicus*) grazing intensities: mob- (stocking densities from 32,000 to 67,000 kg ha^{-1}; duration—24 hr); rotation (stocking density—2500 kg ha^{-1}; duration—35 d); and non-grazed systems were compared based on forage utilization and changes to western snowberry (*Symphoricarpos occidentalis*) (WS) patch volume in a 2-year South Dakota study. Pre- and post-grazing forage height was measured every 2.5 m along multiple 50-m transects with WS patch volume measured every 5 m. Forage utilization (consumed and trampled) ranged from 42 to 90% in mob-grazed areas, and harvest efficiency (forage consumed) ranged from 15 to 64%. WS patch volumes decreased by ≥45% in mob-grazed treatments compared with no change in rotational-grazing and increased cover in non-grazed areas. WS pre-graze patch size influenced mob-grazing impact; patches >6500 cm^3 were browsed or trampled to a greater extent than smaller patches.

Keywords: pasture management, grazing intensity, shrub control, forage grass production

1. Introduction

Western snowberry (WS) (*Symphoricarpos occidentalis*) (also known as 'buckbrush') is a perennial, cool-season shrub, native to the Northern Great Plains (NGP) of the United States [1]. It can grow up to 1 m tall and spreads by seeds and rhizomes. WS can form dense monoculture patches ranging from <2 to 200 m in diameter. This woody species can tolerate poor soils, harsh temperatures, flooding, and drought [2].

Some patches of WS are desirable as thickets provide nesting habitat for ground-dwelling birds, as well as some protection for newborn calves (Charlie Totton, rancher, personal communication, June 2013). Therefore, complete elimination of WS plants in most pastures is not the ultimate management goal. But, over time, uncontrolled patches of this less palatable [3], woody groundcover can reduce plant species diversity and amounts of desirable forage; alter nutrient cycling [4]; and result in economic loss [5, 6].

Options for rangeland perennial weed control vary in implementation and effectiveness and require multiple years of maintenance. After WS removal, biomass of grasses and forbs can increase dramatically [7], although WS densities can rebound in less than a year if control measures cease [8]. Herbicides applied in June resulted in good (64%) to excellent (99%) WS control during the growing season, depending on herbicide and application rate [9, 10] with control in subsequent years ranging from none to excellent. Even with excellent control, herbicides often are more expensive than short or mid-term returns justify [11]. However, because WS occurs in patches, uniform treatment of entire pastures may not be necessary, thus reducing costs and environmental impacts.

Physical techniques, based on timing, alone or combined with grazing are other options for WS management. While a single growing season of mowing did not control WS [10], two mowing events over 3 years reduced WS patch size [1], and increased succulent sprout growth, making the plant more palatable to livestock. Grazing WS patches in early season (May) or left untreated had lower WS densities the following year compared to areas grazed in August [12] or burned with late season fire [8]. Prescribed fires from mid- to late-May combined with goat (*Capra aegagrus hircus*) grazing suppressed WS plants, reduced seed production, canopy cover, and stem density [13, 14] in the NGP. However, goats are not commonly reared in the NGP for a variety of reasons [15].

Cattle grazing for weed control is a natural fit for NGP pastures and rangelands. However, weed management using cattle often has limited success. First and foremost, cattle are expensive to raise and replace and, depending on weed species, may result in problems with nutrition [16], reproduction, toxicity, or have other negative impacts (e.g. off flavors of meat) [2, 17]. Since cattle avoid dung-soiled pasture, selective grazing can occur when stocking rates are low or moderate [18, 19]. Thus, only the most palatable plants are grazed, leading to overgrazing desirable species, and ultimately changing the plant community [20]. In addition, cattle hooves break up sod, leaving areas vulnerable to weed invasion, which is counter-productive to control [5].

Deliberately managing and manipulating cattle stocking rate and density, grazing duration, and seasonal timing based on pasture conditions can promote weed management success [16]. Livestock can consume and/or trample plants and improve pasture nutrient condition and competitiveness of desirable plants through incorporation of manure and urine [5], often with fewer adverse effects on non-target species than herbicide applications. Grazing should occur when the weed is most palatable, vulnerable to injury, and not toxic to the animal.

Mob grazing (or nonselective grazing) using cattle has been promoted as a system to improve soil health and plant conditions [21–23]. This system attempts to mimic animal/vegetation interactions of historic prairie ecosystems, where herds of large herbivores move continuously to new areas as forage is depleted [24]. Although not strictly defined, this system uses dense groups of animals (e.g. >28,000 kg cattle ha^{-1}) in small paddocks for short time periods (typically ≤24 hr. per paddock). Positive attributes include high vegetation utilization [25]; limited selectivity or avoidance of less desirable and potentially weedy plants [26]; and increased trampling of unconsumed forage, manure, and urine [27], which incorporates nutrients and organic matter into the soil [28] and may ultimately lead to higher forage productivity [21]. Some ranchers have adopted this intense system to reclaim specific pastures, following the recommended guidelines to graze for only a few weeks during the season with movement every 12–24 hr. Grazing return periods are often 1 year or longer, or these pastures may be returned to rotational grazing after recovery.

To date, no studies have compared WS plant response to high cattle stocking intensity (mob grazed areas) with low intensity (rotational grazing) or ungrazed treatments. We quantified the impact of mob grazing on WS populations and compared the response with two less-intensive management systems at South Dakota locations. Due to the expense, need for many animals, and labor and time involved to move cattle frequently, the studies were done in cooperation with ranchers who incorporated mob grazing and rotational grazing techniques into their cattle operations. The objective of this study was to quantify the effect of mob-grazed cattle compared with rotational grazed cattle or no grazing on WS size and forage utilization.

2. Methods

2.1. Study site description and treatments

Forage utilization and WS (*Symphoricarpos occidentalis*) data were collected at two South Dakota locations, Chamberlain (southcentral SD; 43.8°N, 99.3°W) and Selby (northcentral SD; 45.3°N, 99.8°W). South Dakota, located in the NGP of the United States, has a continental climate, i.e. cold winter temperatures with snow, and moderate to warm summer temperatures. Most annual precipitation occurs in spring and summer.

At Chamberlain, forage and WS response was quantified in mob-grazed pastures (2013 and 2014), ungrazed pastures (2013) and an early (May through mid-June) rotational-grazed pasture (2014). In Selby, treatments were performed in mob-grazed (2013 and 2014), rotational-grazed (2013), and ungrazed pastures (2014). Pasture vegetation and soil types for each site are listed in **Table 1**. Climate, grazing, and sampling information for the two-year period are provided in **Table 2**. Growing degree days (GDD; base 0°C) for the growing season (March through September) were near (±5%) the 30-yr average (1980–2010) at each year and site (**Table 2**). Precipitation (January through September) was 8% lower than their respective 30-yr averages for both sites in 2013, and 4% lower at Chamberlain and 16.5% lower at Selby, in 2014. Specific GDD and precipitation amounts for sampling dates are reported in Myer [29].

Local producers determined stocking intensity, grazing dates, and paddock size, with cattle moved in mob grazed areas after 24-hr (**Table 2**). The rotational and ungrazed treatments differed among years at the locations due to cattle needs and pasture condition. At Chamberlain

Chamberlain		Selby	
Common name	**Scientific name**	**Common name**	**Scientific name**
Western wheatgrass	*Pascopyrum smithii*	Western wheatgrass	*Pascopyrum smithii*
Smooth brome	*Bromus inermis*	Green needlegrass	*Nassella viridula*
sweet clover	*Melilotus officinalis*	Western snowberry	*Symphoricarpos occidentalis*
Western snowberry	*Symphoricarpos occidentalis*	smooth brome	*Bromus inermis*
red clover	*Trifolium pratense*	Kentucky bluegrass	*Poa pratensis*
Kentucky bluegrass	*Poa pratensis*	Scurfpea	*Psoralidium sp.*
dandelion	*Taraxacum officinale*	Sweet clover	*Melilotus officinalis*
Common sunflower	*Helianthus annuus*		
Musk thistle	*Carduus nutans*		
Common ragweed	*Ambrosia artemisiifolia*		
Bull thistle	*Cirsium vulgare*		
Milkweed	*Asclepias sp.*		
Green needlegrass	*Nassella viridula*		
Needle and thread	*Hesperostipa comata*		
Big bluestem	*Andropogon gerardii*		
Porcupine grass	*Schizachyrium scoparium*		
Sideoats grama	*Bouteloua curtipendula*		
Blue grama	*Bouteloua gracilis*		
Soil types			
Sansarc-Opal clay		Opal-Sansarc clay	
McClure silt loam		Bearpaw-Gettys complex	
Bullcreek clay		Highmore-Bearpaw silt loam/clay loam	
Uly silt loam		Gettys clay loam	

Table 1. Plant species and soil types at Chamberlain and Selby, SD in 2013 and 2014.

mob grazed pastures were mob grazed every-other year, so the 2013 mob grazed pasture was ungrazed in 2012, and the 2014 mob grazed pasture was ungrazed in 2013. Meanwhile, the 2013, ungrazed pasture at Chamberlain was rotationally grazed at a stocking density of 250 kg ha^{-1} for approximately 30 days on 300 ha in 2012, and this same pasture was rotationally grazed in 2014. At Selby, both the mob grazed and rotationally grazed pastures were managed similarly in previous years as the experimental years, and the 2014 ungrazed pasture was rotationally grazed in 2013.

Site		Climate				Grazing information				Sample date	
		GDD Jan - Sept	30 yr avg[a]	Precip. Jan - Sept	30 yr avg[a]	Treatment	Stocking density	Paddock size	Graze duration	pre	post
		Base 0C		mm			Kg/ha	ha			
Chamb.	2013	3262	3416 (-8%)	468	507 (-8%)	Mob	67,000	5	24-hr	19-Jun	8-Aug
						Ungrazed[b]	---	300	---	19-Jun	8-Aug
	2014	3230	3416 (-5.4%)	489	507 (-4%)	Mob	43,800	2	24-hr	9-Jul	17-Jul
						Rotation[c]	250	300	35 days	9-Jul[d]	13-Sep
Selby	2013	2904	3027 (-4%)	366	399 (-8%)	Mob	32,200	2	24-hr	21-Sep	28-Sep
						Rotation	2,600	80	36 days	18-Jun	19-Jul
	2014	2852	3027 (-5.7%)	334	399 (-16.5%)	Mob	37,500	2	24-hr	23-Jul	7-Aug
						Ungrazed	-----	80	---	20-May	6-Sep

[a]30 yr average is based on 1980 to 2010 data for the nearest weather stations to the study site (Chamberlain and Hoven, SD, respectively.

[b]Rotationally grazed in 2012

[c]Grazed in May – mid June

[d]Both sampling dates occured after rotational grazing

Table 2. Climate, grazing information, and sampling dates for 2013 and 2014 at Chamberlain (Chamb.) and Selby, SD locations.

2.2. Vegetative data collection

Three parallel 50-m transects were set up about 10 m apart (pre-graze measurement) immediately prior to the first sampling date (**Table 2**; **Figure 1A**) with two paddocks of each grazing treatment sampled each year (six transects) per site. Average standing forage height was measured to the nearest cm every 2.5 m along each transect, with GPS points recorded (Garmin eTrex 20, Garmin International, Inc., Olathe, KS). Every 5 m along each transect, the closest WS plant was identified, tagged with a metal loop near the plant base, and height (highest point from soil surface) and two perpendicular widths were measured.

After grazing, or in the fall for the non-summer grazed paddocks, transects were reestablished. Average standing forage height, including WS, at each sampling point was measured again and percentage of newly trampled forage (e.g. vegetation remaining that was ≤45° from upright position) was estimated in mob-grazed paddocks only. Tagged WS plants were measured as in pre-grazing, and post-grazing condition (e.g. intact, trampled, browsed) was recorded.

Figure 1. Pasture condition before (A) and after (B) a 24-hr mob grazing event at Chamberlain, SD.

Forage productivity can be estimated using the grazing stick method:

$$\text{Forage productivity} = (\text{average standing forage height} - 10\ \text{cm}) \times 79\ \text{kg ha}^{-1}\ \text{cm}^{-1}. \quad (1)$$

which is the conversion value for a cool season, mixed species pasture with about 90% cover [30]. The 10 cm is subtracted from the height to account for remaining leaf and stubble after grazing. In preliminary data sets, Myer compared grazing stick method to clipping forage biomass at >40 sampling points and found these two estimates were within 15% of each other [29].

Therefore, due to time and labor constraints, grazing stick measurements that accounted for height and percent cover were used to describe relative forage productivity and grazing impact.

2.3. Data analysis

Forage consumption (efficiency) percentage [31] was estimated by:

Forage consumption = [(pre-graze biomass) – (post-graze biomass + trampled)]/(pre-graze biomass) × 100; with biomass estimated using the grazing stick method described above, and assuming the difference in standing forage was consumed by the livestock and not by insects, wildlife, or rodents [32]. Since there was no trampled forage estimate for rotational-grazed paddocks, this was only calculated for the mob-grazed treatment. Additionally, forage utilization (consumed + trampled forage) was estimated at each sampling point and was based on change in biomass (estimated using the grazing stick method), including the trampled forage. Pre- and post-grazing WS relative plant volume was estimated for each tagged plant using the equation:

WS volume = height × width 1 × width 2; with height measured at the highest point on the plant from the soil surface, width 1 as the widest horizontal measure of the WS, and width 2 the width of the WS perpendicular to the width 1 measurement.

Matched paired one-tailed (post-graze<pre-graze) t-tests were used to compare pre- versus post-grazing WS plant volume and estimated forage biomass at each point along the transect at a significance va1lue of $P \leq 0.10$. Data were combined when appropriate. Data from ungrazed pastures were examined with a one-tailed matched paired analysis test with the assumption that spring forage < fall forage.

Binomial analysis of WS plant volume data (yes = less volume post grazing (or in the fall for non-summer grazed treatment); no = same or greater volume) using the equation: [33]

$$\left[p \pm t_{(0.1)}\, \text{sqrt}\,(p\,(1-p)/n)\right] \tag{2}$$

was used to determine if grazing intensity treatments impacted individual WS plant volume. In addition, WS plants were separated into two volume classes (<6500 cm^3 and >6500 cm^3) based on the median WS plant size in grazed pastures and analyzed using two-tailed matched paired t-tests to determine if pre-graze volume impacted cattle interaction with plants.

3. Results

3.1. Chamberlain

Estimated forage biomass before mob grazing was 6100 and 2840 kg ha^{-1} in 2013 and 2014, respectively (**Table 3**). Stocking density was greater and individual paddock size larger in 2013 (67,200 kg ha^{-1} on 5 ha) than 2014 (43,680 kg ha^{-1} on 2 ha). Harvest utilization (consumed + trampled) in mob-grazed areas were similar and >90% each year. Harvest efficiency (amount consumed) was also similar and >60% each year.

Site	Year	Treatment	Standing Forage			Trampled forage	Utilization[a]	Efficiency[b]
			Pre-graze	Post-graze	t test			
			Kg/ha			Kg/ha	%	
Chamb.	2013	Mob	6,100	80	****	2,130	99	64
		Ungrazed[c]	1,190	2,690	****			
	2014	Mob	2,840	160	****	894	94	63
		Rotation[d]	1,980	1,660	*			
Selby	2013	Mob	1,900	1,110	****	505	42	15
		Rotation	2,690	790	****			
	2014	Mob	1,980	240	****	785	88	48
		Ungrazed	470	2,920	****			

***Significant at $p<0.0001$, *significant at $p<0.1$

[a]Utilization = forage consumed and trampled, calculated by [(pre-graze) - (post-graze) / (pre-graze) x

[b]Efficiency = forage consumed, calculated by [(pre-graze) - (post-graze + trampled forage) / (pre-graze)

[c]Rotationally grazed in 2012

[d]Both sampling dates occured after rotational grazing

Table 3. Impact of pasture management on forage at Chamberlain (Chamb.) and Selby, SD experimental sites in 2013 and 2014.

The comparison pasture was not grazed in 2013 but had been rotationally grazed in 2012. On June 19, forage biomass was estimated at 1190 kg ha^{-1}, whereas on August 8, biomass increased to 2690 kg ha^{-1} (**Table 3**). Between the first and second sampling there was >200 mm of rainfall. In 2014, the comparison pasture was grazed at a stocking rate of 250 kg ha^{-1} from May to mid-June, which was prior to the first sampling. Forage biomass on July 9 and September 13 was similar ($P = 0.1$), averaging about 1800 kg ha^{-1}. The apparent lack of growth may be explained by dormancy of the dominant cool season species, lack of rainfall (<12 cm) between sampling dates, and a grasshopper (*Caelifera* sp.) infestation that consumed forage regrowth.

Response of WS plants to mob grazing was similar in both years, with data combined over years. About 95% (±4%) of the measured plants were reduced post-grazing by an average of 63% (**Figure 2A**). Forage near WS plants was consumed (about 75% less biomass present), rather than trampled, and WS appeared to be browsed (stems and leaves removed). WS plant response in the 2013 ungrazed pasture indicated no difference in WS plant volume between the first and second sampling ($P = 0.43$). In 2014, WS plants were reduced in volume by about 19% (from 8850 to 7050 cm³) between the first and second sampling dates ($P = 0.01$) even though grazing occurred prior to the first sampling. This was due to a large grasshopper infestation in the area.

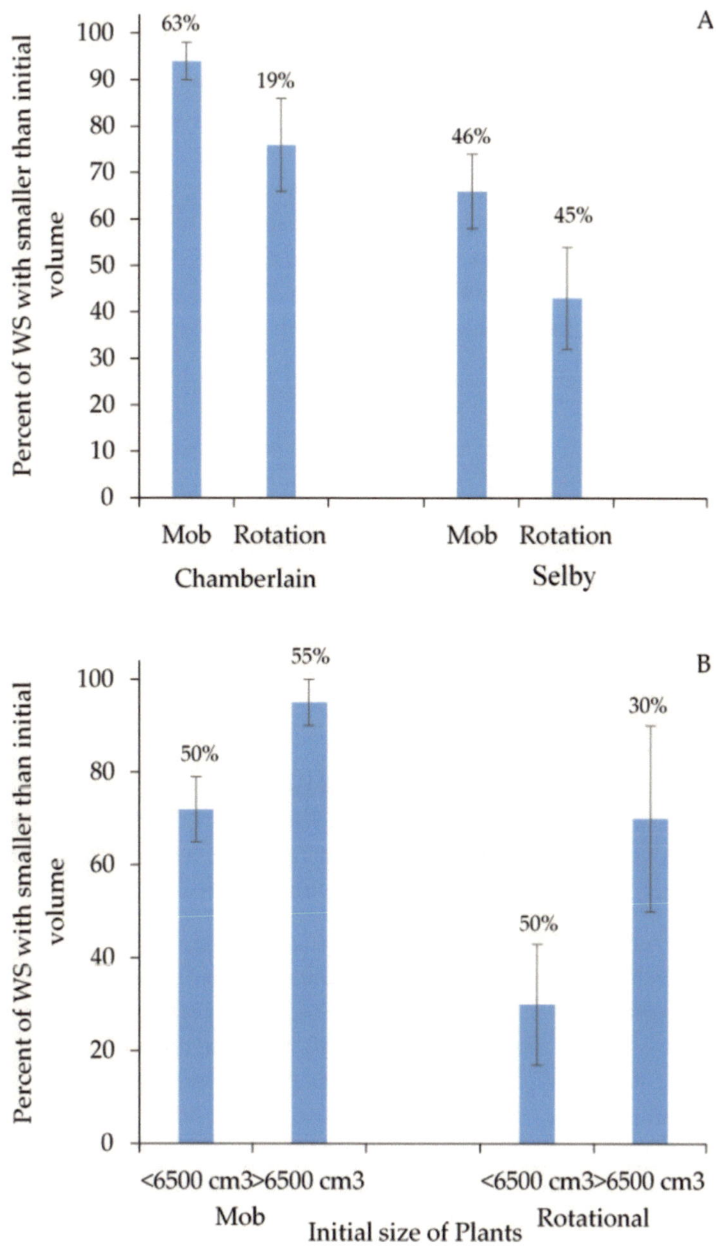

Figure 2. Percent (±SE) of western snowberry with smaller volume post-graze by grazing system (A), and percent (±SE) of western snowberry with small volume post-graze based on initial size (<6500 cm³ and >6500 cm³) by grazing system (B). Numbers above bars represent the average volume reduction of western snowberry plants in the respective category.

3.2. Selby

Forage at Selby averaged about 1940 kg ha^{-1} each year prior to mob grazing (**Table 3**). After the 24-hr grazing event, forage remaining was 1110 kg ha^{-1} in 2013 and 240 kg ha^{-1} in 2014. Forage consumption and utilization were estimated at 15 and 42%, respectively, in 2013. In 2014, efficiency and utilization were estimated at 48 and 88%, respectively. The three-fold increase in forage consumption (efficiency) in 2014 compared to 2013 may have been due to timing of the grazing. Forage was likely more mature and less palatable for cattle in late September (2013) compared to late July (2014). The increase in both consumption and utilization may have also been due to the slightly higher stocking density in 2014 compared to 2013 (**Table 2**). In the 2013 rotation-grazed pasture, the forage biomass averaged 2690 kg ha^{-1} pre-graze and about 790 kg ha^{-1} post-graze, with an estimated 70% consumption and utilization, as very little newly trampled biomass was present. In 2014, the ungrazed comparison pasture had about 470 and 2920 kg ha^{-1} at the first and second sampling, respectively.

Volume data from WS plants were combined for the 2013 and 2014 mob grazing treatment, with 66% (±8%) of the tagged plants decreasing in volume by 46% after grazing. In the rotational-grazed area, pre- and post-sampling volumes were similar and averaged 15,000 cm^3. However, 43% of these sampled plants had a 45% reduction in volume, but the remaining plants increased in volume by about 90%. Basal stem counts (data not shown) indicated that WS plants in mob-grazed areas had fewer stems ($P = 0.001$) after grazing, whereas no difference in stem number was observed in rotational-grazed plots. In the 2014 ungrazed pasture, 74% of the tagged plants increased in volume by an average of 5000 cm^3, a 3000% increase from the first to the second sampling.

3.3. Initial WS plant volume and grazing impact

Initial WS plant volume impacted final volume after mob grazing. Mob grazing data, combined by location, indicated that the median plant size was about 6500 cm^3. When initial plant volume < 6500 cm^3, 73% (±7%) of these plants had a 42% reduction in volume. However, about 87% (±5%) of the larger plants were reduced in volume by about 62%.

In the early spring rotationally-grazed paddock at Chamberlain (2014), initial volume did not impact final size ($P = 0.46$). About 66% of all plants increased in size an average of 168% (±81%). In the late-season rotation treatment at Selby in 2013, about 50% (±13%) of the small plants were reduced in volume by about 52% (**Figure 2B**), with the remaining plants increasing in volume an average of 150%. About 44% of the large plants were reduced in volume by about 38%, with the remaining plants increasing an average of 37%. While the plant size reduction in the less intensively grazed rotational treatment was similar between the large and small plant classes ($P = 0.46$), the increase in size of the small plants was greater than the size increase of the large plants ($P = 0.02$). This may be due to smaller plants being trampled and stems spread apart thereby increasing the final volume (i.e. plants lost vertical height but both horizontal lengths increased), whereas larger plants may have been more difficult to trample.

These data indicate that WS plants were more impacted by mob grazing compared with plants in paddocks rotationally-grazed early or later in the season. Larger plants in mob-grazed areas tended to be more damaged than smaller plants.

4. Discussion

Cattle in NGP mob-grazed settings were more competitive for available forage, and were less selective in consumption, eating vegetation that would normally be avoided in a less intense grazing. The high stocking densities also resulted in more trampling and greater animal impact (e.g. dung deposition, data not shown) per unit area [29]. Other studies have reported similar results in other intensive-grazing systems although terminology [e.g. ultra-high stocking density [23]; intensive stocking [34]; cell-grazing [35]; high intensity, low frequency grazing [36], stocking rates, grazing duration, and seasonal timing often differ. High stocking densities have been shown to maintain animal performance if carefully managed [36]. Lush regrowth during the rest period following an intense grazing event increased forage crude protein (from 8.9 to 10.2%) and digestibility (from 44.6 to 54.7%) compared with more mature forage in less-intensively grazed areas [36]. Timing of grazing events, both within and among seasons on the same parcels, must be carefully controlled as repeated grazing when grass is at a vulnerable growth stage can result in rangeland degradation [37, 38].

Other studies have reported that cattle graze less palatable, weedy species when grazing intensity is high. For example, cattle have browsed prickly pear (*Opuntia macrorhiza*) [39], absinth wormwood [29], and thistles [40], species that are typically avoided in low-intensity grazing. The least desirable species at Australian sites, purple wiregrass (*Aristida ramose*) and gray tussock-grass (*Poa sieberiana*), decreased 45% in basal diameter in a cell-grazing treatment with a stocking rate of about 35,000 kg ha^{-1} and moved every 1–3 days compared with <5% decreases observed in continuously grazed sites [35]. These results suggest that during mob-grazing events, animals will browse less desirable species. In addition, mob-grazing, or similar high stocking-density, low frequency grazing management, has been suggested to maximize forage use [21], aid in maintaining a balance of desirable and undesirable vegetation [41] and may enhance nutrient cycling in the paddock with minimal to no risk to animal gains if properly managed [42, 43]. However, mob-grazing should be strictly managed with recovery periods for forage regrowth to ensure adequate feed. Returns to management can be low for mob-grazing [45] if high stocking densities for long periods reduce average daily gain per animal [46] and may degrade range resources and resilience.

Size of WS plants influenced the efficacy of mob-grazing for weed management. In contrast to absinth wormwood (*Artemisia absinthium*) (AW) where small patches and plants were most affected by mob-grazing [44], larger WS plants were most impacted. Larger WS plants may have leaves closer to the cattle's face, which may facilitate browsing strictly due to convenience, even though the stems are woody. Smaller AW plants, which have herbaceous rather than stiff woody stems, may be more easily trampled and/or consumed.

5. Conclusions

Mob-grazing with cattle reduced forage selectivity and utilized undesirable plants compared to low stocking density rotational grazing. Long-term benefits of mob-grazing, while difficult to quantify in short-term studies, can be positive and numerous SD ranchers have adopted this technique to their advantage. In this study, we realize that stricter control of variables such as stocking density, timing, and pasture size, may have resulted in more repeatable and statistically significant results. However, this research was conducted on working ranches and represents actual producer management decisions based on forage pasture conditions, annual climate, and cattle needs. Therefore, the results may be more applicable to NGP ranchers. Ranchers who are interested in using mob grazing should start small to determine how best to employ this system in their operation. Future research that combines mob-grazing at the most vulnerable stages of weed species growth with other management practices (e.g. herbicide application or pasture fertilization) should be considered. We conclude that mob-grazing can decrease forage selectivity and be a useful tool in for integrated weed management of WS, especially for plants larger than 6500 cm^3 in the NGP.

Acknowledgements

Thanks to Mr. L. Perman, Selby, SD; Mr. D. Seick, Selby, SD; and Mr. C. Totton, Chamberlain, SD for providing access to land and cattle. Funding support for this study was provided by the South Dakota Agricultural Experiment Station and USDA, NRCS Conservation Innovation Grant (CIG) 3FH560 "Demonstrating Mob Grazing Impacts in the Northern Great Plains on Grazing Land Efficiency, Botanical Composition, Soil Quality, and Ranch Economics."

Author details

Heidi Reed[1*], Sharon Clay[1], Alexander Smart[2], David Clay[1] and Michelle Ohrtman[1]

*Address all correspondence to: hxm5183@psu.edu

1 Department of Agronomy, Horticulture and Plant Science, South Dakota State University, Brookings, South Dakota, United States

2 Department of Natural Resource Management, South Dakota State University, Brookings, South Dakota, United States

References

[1] Manske LL. Western snowberry biology [Internet]. 2006. Available from: www.ag.ndsu.edu/archive/dickinso/research/2005/PDF/range05a.pdf [Accessed: 2017-07-01]

[2] DiTomaso JM. Invasive weeds in rangelands: Species, impacts, and management. Weed Science. 2000;**48**:255-265

[3] Pelton J. Studies of the life-history of *Symphoricarpos occidentalis* hook in Minnesota. Ecological Monographs. 1953;**23**:17-39

[4] Pimentel D, Zuniga R, Morrison D. Update on the environmental and economic costs associated with alien-invasive species in the United States. Ecological Economics. 2005; **52**:273-288

[5] Popay I, Field R. Grazing animals as weed control agents. Weed Technology. 1996;**10**: 217-231

[6] Frid L, Hanna D, Korb N, Baver B, Bryan K, Martin B, et al. Evaluating alternative weed management strategies for three Montana landscapes. Invasive Plant Science and Management. 2013;**6**:48-59

[7] Bowes GG, Spurr DT. Improved forage production following Western snowberry (*Symphoricarpos Occidentalis* hook.) control with Metsulfuron methyl. Canadian Journal of Plant Science. 1995;**75**:935-940

[8] Romo JT, Grilz PL, Redmann RE, Driver EA. Standing crop, biomass allocations patterns and soil-plant water relations in *Symphoricarpos occidentalis* hook. Following autumn or spring burning. American Midland Naturalist. 1993;**130**:106-115

[9] Sbatella GM, Wilson RG, Sleugh B. Western snowberry (*Symphoricarpos occidentalis*) control with aminopyralid and metsulfuron. Weed Technology. 2011;**25**:616-619

[10] McCarty MK. Control of Western snowberry in Nebraska. Weeds. 1967;**15**:130-133

[11] Masters RA, Sheley RL. Principles and practices for managing rangeland invasive plants. Journal of Range Management. 2001;**54**:502-517

[12] Bailey AW, Irving BD, Fitzgerald RD. Regeneration of woody species following burning and grazing in Aspen Parkland. Journal of Range Management. 1990;**43**:212-215

[13] Smart AJ, Daniel J, Bruns K, Held J. Browsing of Western snowberry by goats and sheep. Sheep and Goat Research Journal. 2006;**21**:1-5

[14] Smart AJ, Troelstup NH Jr, Bruns KW, Daniel JA, Held JE. Western snowberry response to fire and goat browsing. Sheep and Goat Research Journal. 2007;**22**:20-25

[15] Sahlu T, Dawson LJ, Gipson TA, Hart SP, Merkel RC, Puchala R, et al. ASAS centennial paper: Impact of animal science research on United States goat production and predictions for the future. Journal of Animal Science. 2009;**87**:400-418

[16] Frost RA, Launchbaugh KL. Prescription grazing for rangeland weed management: A new look at an old tool. Rangelands. 2003;**25**:43-47

[17] Fishel F. Plants poisonous to livestock. Bulletin G-4970 University of Missouri Extension

[18] MacDiarmid BN. Cattle dung patch 3. Distribution and rate of decay of dung patches and their influence on grazing behavior. Journal of the British Grassland Society. 1972;**27**:48-54

[19] Marten GC, Donker JD. Selective grazing induced by animal excreta I. Evidence of occurrence and superficial remedy. Journal of Dairy Science. 1964;**47**:773-776

[20] Anderson VJ, Briske DD. Herbivore-induced species replacement in grasslands: Is it driven by herbivory tolerance or avoidance? Ecological Applications. 1995;**5**:1014-1024

[21] Savory A, Butterfield J. Holistic Management: A Commonsense Revolution to Restore our Environment. 3rd ed. Washington, DC: Island Press; 2016. 552 p. ISBN: 9781610917438

[22] Gordon, K. Mob Grazing 101 [Internet]. 2011. Available from: hereford.org/static/files/0111_MobGrazing.pdf [Accessed 2017-05-01]

[23] Hafla AN, Soder KJ, Hautau M, Rubano MD, Mayer B, Stout R. Case study: Dairies using self-described ultra-high stocking density grazing in Pennsylvania and New York. The Professional Animal Scientists. 2014;**30**:366-374

[24] Fuhlendorf SD, Engle DM. Restoring heterogeneity on rangelands: Ecosystem management based on evolutionary grazing patterns. Bio Science. 2001;**51**:625-632

[25] Barnes MK, Norton BE, Maeno M, Malechek JC. Paddock size and stocking density affect spatial heterogeneity of grazing. Rangeland Ecology and Management. 2008;**61**:380-388. DOI: 10.2111/06-155.1

[26] Bailey DW, Brown JR. Rotational grazing systems and livestock grazing behavior in shrub-dominated semi-arid and arid rangelands. Rangeland Ecology and Management. 2011;**64**:1-9

[27] Pleasants AB, Shorten PR, Wake GC. The distribution of urine deposited on a pasture from grazing animals. Journal Agricultural Science. 2006;**145**:81-86

[28] Frank DA, McNaughton SJ. Evidence for the promotion of aboveground grassland production by native large herbivores in Yellowstone National Park. Oecologia. 1993;**96**: 157-161

[29] Myer H. Mob grazing as a perennial weed management tool in South Dakota grazing lands [thesis]. Brookings, SD: South Dakota State University; 2015

[30] NRCS Grazing or Pasture Stick: Instructions for Using [Internet]. 2013. Available from: file:///C:/Users/hxm5183/Downloads/PASTURE_STICK4.pdf [Accessed: 2018-04-01]

[31] Heady HF, Child RD. Rangeland Ecology and Management. Boulder, CO: Westview Press; 1994. p. 544

[32] Hewitt GB, Burleson WH, Onsager JA. Forage losses caused by the grasshopper *Aulocara ellotti* on shortgrass rangeland. Journal of Range Management. 1976;**29**:376-380

[33] Steel RGD, Torrie JH. Principles and Procedures of Statistics. 2nd ed. New York: McGraw-Hill, Inc.; 1980. ISBN 0-07-060926-8

[34] Brock BL, Evans L, Owensby CE. Spatial and temporal grazing patterns on intensive-early stocked and season-long stocked pastures. Crop Forage and Turfgrass Management. 2017;**3**. DOI: 10.2134/cftm2016.00016

[35] Earl JM, Jones CE. The need for a new approach to grazing management—Is cell grazing the answer? Rangeland Journal. 1996;**18**:327-350

[36] Taylor JCA, Kothmann MM, Merrill LB, Elledge D. Selection by cattle under high-intensity low-frequency, short duration, and Merrill grazing systems. Journal of Range Management. 1980;**33**:428-434

[37] Ellison L. Influence of grazing on plant succession of rangelands. The Botanical Review. 1960;**26**:1-78

[38] Manske LL. General description of grass growth and development and defoliation resistance mechanisms. NDSU Dickinson Research & Extension Center: Range Management Report DREC. 1998;**98-1022**:12

[39] DeLoach CJ. Past successes and current prospects in biological control of weeds in the United States and Canada. Natural Areas Journal. 1991;**11**:129-142

[40] DeBruijn SL, Bork EW. Biological control of Canada thistle in temperate pastures using high density rotational cattle grazing. Biological Control. 2006;**36**:305-315

[41] Taylor JCA, Brooks TD, Garza NE. Effects of short duration and high intensity, low-frequency grazing systems on forage production and consumption. Journal of Range Management. 1993;**46**:118-121

[42] Biondini ME, Manske L. Grazing frequency and ecosystem processes in a Northern Mixed Prairie, USA. Ecological Applications. 1996;**6**:239-256

[43] Taylor CA, Kothmann MM, Merrill LB, Elledge D. Selection by cattle under high-intensity low frequency, short duration, and Merrill grazing sytsems. Journal of Range Management. 1980;**33**:428-434

[44] Myer H, Smart A, Ohrtman M, Clay D, Clay SA. Mob vs rotational grazing: Impact on forage use and absinth wormwood (*Artemisia absinthium*). In: Forages. InTech; 2018. (in review)

[45] Janssen L, McMurtry B, Stockton M, Smart A, Clay S. An economic analysis of high-intensity, short-duration grazing systems in South Dakota and Nebraska. In: Proceedings of the Agricultural & Applied Economics Association; 26-28 July 2015; San Francisco. 2015. pp. 2-28

[46] Smart AJ, Derner JD, Hendrickson JR, Gillen RL, Dunn BH, Mousel EM, et al. Effect of grazing pressure on efficiency of grazing on North American Great Plains rangelands. Rangeland Ecology and Management. 2010;**63**:397-406

Chapter 3

Forage Yield Estimation with a Process-Based Simulation Model

James R. Kiniry, Sumin Kim, M. Norman Meki and Mari-Vaughn V. Johnson

Additional information is available at the end of the chapter

http://dx.doi.org/10.5772/intechopen.79987

Abstract

A process-based simulation model of natural grasslands and improved pastures can be used to compare mean productivity and stability of forage productivity across years, agroecological regions, and management approaches. Model simulations can help farmers develop management practices to optimize livestock stocking rates and nutrient management for native and improved grasses on different soils with varying rainfall amounts. Likewise, forages are adapted to a wide variety of soils, rainfall zones, and latitudes. The objective of this chapter is to describe the Agricultural Land Management Alternative with Numerical Assessment Criteria (ALMANAC) model that simulates a wide variety of environmental and management impacts on forage production, soil health, and conservation concerns, including nutrient and sediment losses. We describe the various processes simulated in the model and input data requirements. We also describe how to derive plant parameters for various forage plant species. The model has been applied to simulate forage yields across years and diverse environments in the U.S. and tested using published forage yield data from Natural Resources Conservation Service, United States Dept. of Agric. Many common native and introduced grasses or grass mixtures in the U.S. have been successfully simulated. We also describe and discuss knowledge gaps for the model that future research should address to improve this and similar simulation models.

Keywords: simulation modeling, native grasses, improved grasses, environmental quality simulation, forage management

1. Introduction

A process-based simulation model of natural grasslands and improved pastures offers managers a science-based decision tool with many possible applications. Such a model can be used to compare mean value of and stability of forage productivity across years, agroecological regions, and management approaches. Model simulations can help farmers develop practices to best manage livestock and to most effectively fertilize pastures for native and improved grasses on different soils with varying rainfall amounts. Likewise, how forage productivity responds to different grazing management can be simulated. The ideal model for these applications would have sufficient detail to simulate several plant species, soils, and climatic conditions without excessive input requirements. The model should be able to simulate forage legumes, improved grasses, and common native grasses, as sole crops or mixtures. The required plant parameters should be readily derived from published studies in conjunction with measurements that can be obtained without an inordinate amount of time and effort given to field experiments. This model will have process-based components to simulate leaf area growth, biomass production, and nutrient uptake. In addition, the soils and weather data needed should be readily available, and there should be data sets with sufficient detail for validating forage production simulations. Historically, there has been a diversity of process-based models developed to simulate dynamics of grass growth and dry matter yield for different species. Some examples of process-based models are GRASIM [1, 2], Simulation of Production and Utilization on Rangelands (SPUR) model [3–5], the Ecosystem Level Model (ELM) [6], and DAFOSYM [7].

GRASIM is a grazing simulation model designed to simulate intensive rotational grazing management linked to components (e.g., carbon, nitrogen, and water budgets) of the pasture system This model predicts daily growth rate, biomass accumulation, protein and fiber content, and water and nutrient levels [1]. Simulation of Production and Utilization of Rangelands (SPUR) is a mechanistic process model designed to simulate growth initiation, germination, carbon assimilation, translocation between roots and shoots, N mineralization, and nitrogen uptake [8, 9]. SPUR has been modified and incorporated into the Integrated Farm System Model, IFSM, to simulate the growth and competition of multiple plant species in pastures [10]. The Dairy Forage System Model, DAFOSYM, is a simulation model of the dairy forage system. This model simulates plant growth of lucerne *(Medicago sativa)*, maize (*Zea mays* L.), small grains, and soybean (*Glycine max*) using historical weather data. DAFOSYM evaluates the forage qualities and accumulation of dry matter based on daily weather and soil moisture.

1.1. Constraints on forage production

Forages are adapted to a wide variety of soils, rainfall zones, and latitudes. The environment imposes different constraints on forage production, including but not limited to the following:

1. Varying durations of growing seasons due to temperature and light availability. These are obviously highly dependent on latitude.

2. Available water, including low annual rainfall, variable intra-annual rainfall patterns, and flooding events.

3. Soil attributes:

    a. pH

    b. Soil depth (rooting zone and hydrological dynamics)

c. Soil water storage capacity as impacted by soil texture

d. Slope and rock fragment that impact infiltration/runoff rates

e. Aeration differences with some soils having prolonged flooding

f. Soil nutrient variability, due to either inherent soil fertility or applied nutrients that include organic and inorganic fertilizers

The complexity of these factors and their interactions make forage management a challenge. Animal stocking rates, haying frequency, and optimum fertilizer applications add to the complexity. A producer must take all of these factors into account, plus personal experiences and expert opinion to manage forage lands. However, a process-based simulation model could provide a more science-based approach to management decision making. By systematically simulating forage growth with different soils, weather, plant species, and management systems, such a model will be a valuable tool. In addition, built-in water and nutrient balances in the model allow users to derive guidelines that producers can apply to better adjust management practices among years with high, low, or normal rainfall patterns. Likewise, impacts of various management scenarios on soil erosion, soil organic matter buildup or depletion, and water quality can be evaluated.

The ALMANAC model is adaptable to any species of forage in any geographic location, provided adequate input data are available to inform the model simulation. It can be used to simulate a diversity of forage species as well as woody species, crops, and interspecies competition. The model can simulate plant growth using independently derived plant parameters, with no recalibration among sites. It includes components for the water balance, the nutrient balance, and interception of solar radiation by monocultures or competing plant species. Daily values for weather variables, including temperature and rainfall, are required. Soil inputs are readily available from published USDA-NRCS soil surveys. When fully calibrated, ALMANAC has been shown to reasonably simulate native grass productivity on diverse sites in the U.S. as well as improved grass species at several U.S. sites as discussed below. The objective of this chapter is to describe the Agricultural Land Management Alternative with Numerical Assessment Criteria (ALMANAC) model [11], a process-based model capable of considering a broad variety of environmental and management impacts on forage production, soil health, and conservation concerns, including nutrient and sediment losses.

This chapter consists of five sections:

ALMANAC model simulated processes: The basic processes ALMANAC simulates are described, including processes specific to forage simulation, to include:

a. Cessation of forage plant development due to intense drought stress (Growing degree day accumulation stops temporarily)

b. Midseason dormancy in cool season forages when the daylength gets sufficiently long. This model functionality enables simulation of the bimodal growth pattern observed in such grasses (e.g., tall fescue (*Schedonorus arundinaceus* (Schreb.) Dumort)).

c. Plant competition for water and nutrients, including woody species competing with herbaceous forages

Soils and weather data: The available and required soils and weather data are described.

Deriving plant parameters for a forage species and accommodating ecotypes: The steps for deriving plant parameters for various forage species and ecotypes are outlined.

Model testing against independent data: Soft calibration of the model via comparison of outputs to independent data to ensure the model is working reasonably well is described and discussed.

Knowledge gaps and areas for future improvement as a guide for additional research: Finally, the knowledge gaps and potential areas for improvement are outlined, as a guide for potential additional research.

2. ALMANAC model simulated processes

Phenological development defines the duration of various plant growth stages and determines the length of the forage growing season. The ALMANAC model simulates phenological development with a growing degree day (GDD) system, species-specific base temperature, and optimum growing temperature. The sum of GDD calculates the duration of the growing season. Anthesis date is predicted with a defined fraction of the total GDD sum to physiological maturity. Daylength and drought affect simulated forage phenology as described below.

Leaf area growth is simulated on a whole canopy basis, with potential leaf area index (LAI) defined for each species/ecotype/variety. These are hereafter referred to as just "species." The climate and soils at different sites often dictate the plant density of forages, thereby affecting the potential LAI.

The development of LAI over the growing season is simulated with a 0.0 to 1.0 "S" curve defined for each species. Thus, LAI is simulated as a function of the ratio (current summed GDD)/(GDD to maturity). This ratio typically approaches 1.0 as the plants approach anthesis and transition from forage production to reproduction. The "S" curve thus defines the potential leaf area growth over the growing season.

Daily dry matter accumulation is simulated using a radiation use efficiency (RUE) approach. The potential dry matter produced each day is a function of the amount of photosynthetically active radiation (PAR) intercepted by the leaf canopy on that day. The RUE is a species-specific value (g of dry matter per MJ of intercepted PAR).

Partitioning among plant parts is also on a whole canopy basis. The root and shoot partitioning is defined by two parameters. Plants initially partition a greater fraction of the total dry matter production into the roots. This fraction decreases as plants approach anthesis. Stresses, especially drought, reduce the above-ground dry matter production more than the root dry matter production. This causes drought stress to change the simulated root:shoot ratio.

Partitioning of plant energies to the seed is simulated with a harvest index (HI) approach. The fraction of the total plant weight in the seed at maturity relative to the total plant weight is the species-specific HI parameter. While very small for most forage species due to their relatively small fruits and seeds, the partitioning into the seed begins after anthesis and is complete by physiological maturity.

Environmental stresses decrease leaf area expansion and dry matter accumulation. As described below, the model simulates the impacts of a variety of stresses each day. The most severe stress each day constrains leaf area growth and dry matter accumulation. Leaf area growth is more sensitive, especially to drought, than is dry matter growth.

Drought stress is simulated using the potential evapotranspiration (PET), calculated as a function of daily weather variables. The available soil water in the current rooting zone is calculated each day based on rainfall, soil infiltration, and soil water-holding capacity. If available soil water in the current rooting zone is insufficient to meet the plant's demand (based on PET and leaf area index), the model simulates a drought stress response in the plant through decreased leaf expansion rates and reduced dry matter accumulation rates.

Nutrient stresses, particularly nitrogen (N) and phosphorus (P) stresses, reduce plant growth. These nutrient stresses are simulated with a supply and demand approach. Plant N and P nutrient uptake is simulated with three input parameters that define how nutrient demand changes during the growing season. For each plant species, the optimum amount of available N and P is defined for each species early in plant development, near anthesis, and at physiological maturity. These three values are used to calculate the potential nutrient uptake from the soil each day. If the N and P in the current rooting zone are insufficient to meet demand (calculated from the optimum percentage of the nutrient and the potential daily plant dry matter growth), the model simulates nutrient stress by decreasing the species' dry matter accumulation rate and leaf expansion rate.

Temperature stress can also reduce plant growth in the model. Each plant species has a defined base temperature and an optimum temperature. When daily temperature is below the base temperature, cold temperature stress occurs. When temperatures are above the optimum, high temperature stress occurs.

Aeration stress is also simulated. When soils are saturated with water, aeration stress occurs in the model. Plants have variable sensitivity to aeration stress, as defined by the species-specific value of critical aeration factor (CAF). Plants such as eastern gamagrass (*Tripsacum dactyloides* L.) and rice (*Oryza sativa*) are less sensitive to poor aeration conditions, such as flooding, while upland grasses are more sensitive.

There are components in the model developed specifically for forage simulation. Forage development is not only dependent on GDD accumulation, but also on daylength and stresses. In order to accommodate growth dynamics typical of arid ecosystems where forage species are grown without irrigation, the model was modified so that sufficient drought stress stops GDD accumulation in the model. This is in addition to the direct effects on leaf area growth and dry matter increases as discussed above. As we began simulating plant growth in more arid environments, we had to introduce the ability to halt plant development when drought stress became sufficiently intense. Thus, we introduced a function that stops GDD accumulation (thus stopping phenological development), when the zero-to-one drought stress factor is less than 0.4.

The ALMANAC model is capable of simulating growth patterns exhibited by different types of forages. Cool season forages such as tall fescue often exhibit two intervals of active growth, with a slowdown during the hottest days of the year. Actual growth patterns of tall fescue in southwestern Missouri over 3 years [12, 13] are shown in **Figure 1**. We incorporated a mid-season dormancy function to simulate this. Thus as daylength gets sufficiently long plant growth slows and stops. The model now simulates this rapid growth in the spring, slowing and stoppage of growth near mid-season, and subsequent late summer and early fall growth. The forage-simulation functionality of ALMANAC stops plant growth and development when the maximum photoperiod of the year is reached and restarts growth when the photoperiod subsequently gets sufficiently short to trigger reinitiation of growth. This model functionality was developed based on observed tall fescue growth curves measured in Missouri [12–14].

Subsequently, we tested the model's simulation of tall fescue yields with USDA-NRCS reported yields for a number of sites and soils across the main regions of tall fescue pastures in the U.S. [14]. We used long-term measured weather and the appropriate soil parameters for these simulations. We compared the simulated yields to the reported yields for the low-yielding sites, the high-yielding sites, and for all the sites pooled (**Figure 2**). The model with this function did an excellent job of simulating tall fescue yields on sites with differing reported yields across the main areas of tall fescue production in the U.S.

Additionally, ALMANAC simulations accommodate winter dormancy, typically observed in forage species when daylength gets sufficiently short in the fall. This capacity has been well tested on winter wheat [15], which is planted in the fall, goes dormant during the winter, then restarts growth in the spring. A parameter (DORMNT) defines this interval by defining the hours of photoperiod near the minimum for the latitude when plants are dormant. If the value is 1.0, during the winter when the photoperiod is within 1.0 hour of the minimum for the latitude, plants remain dormant.

Simulation of grazing and hay harvest is especially important when simulating forages. The model resets development (summed GDD, LAI, and height) when the simulation includes a grazing event or the forage is cut for hay. The model simulates a daily value for plant height from the fraction of GDD relative to the physiological maturity value and a species-specific plant height parameter (CHT). When forages are grazed or cut for hay, this height is reduced. If grazing or hay cutting reduces the plant height by 90%, the summed GDD for that day is reduced by 90% and the leaf area and above-ground dry matter is reduced by 90%. The plants then begin regrowth the following day.

Forage plant communities often have mixtures of species, due to the diversity typical of a native prairie, due to intercropping of legumes and grasses to better accommodate nutrient demands, or due to invasion of the forage site by undesirable herbaceous or woody plants. The ALMANAC model is capable of simulating both nitrogen fixation benefits to non–nitrogen-fixing species and competition between plant species. The ALMANAC model was initially

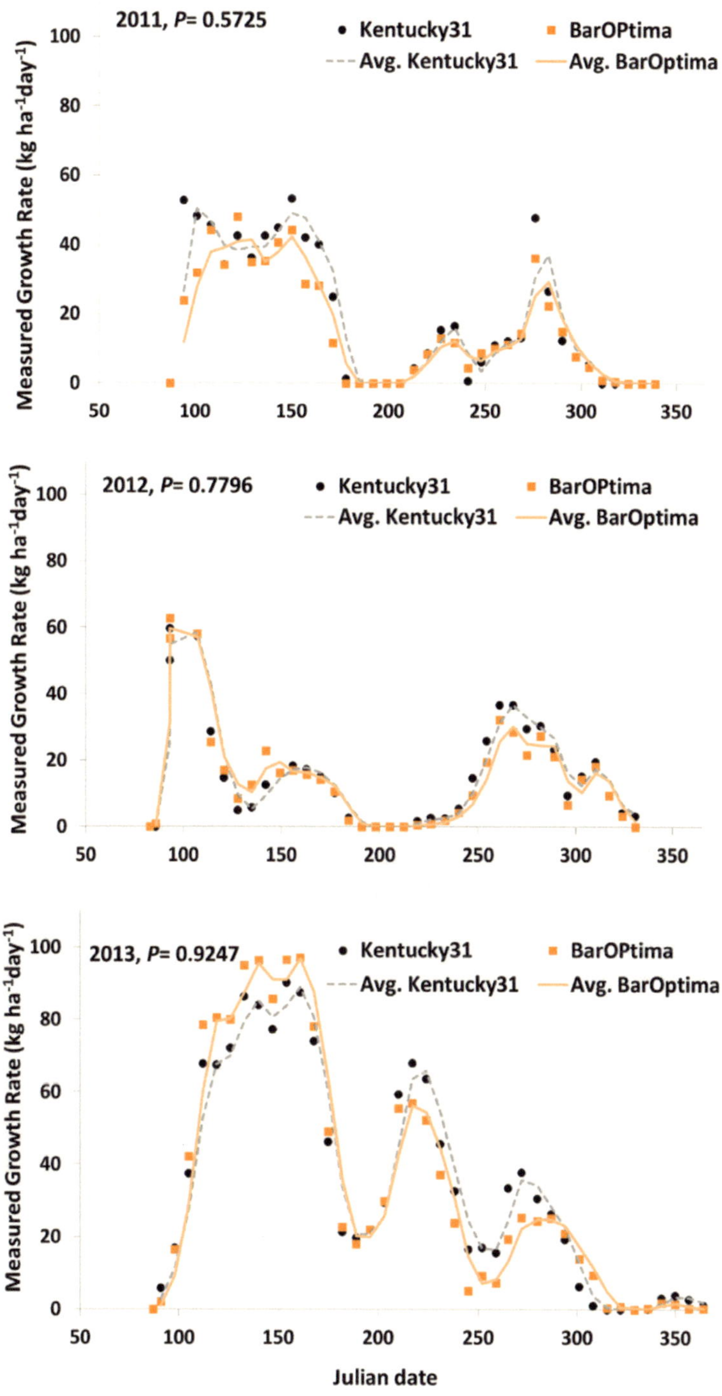

Figure 1. Measured plant growth rates (kg ha^{-1} day^{-1}) for "Kentucky 31" and "Bar Optima" tall fescue in 2011, 2012, and 2013 at Mt. Vernon, MO. The Wilcoxon Rank Sum test was performed to compare "Bar Optima" and "Kentucky 31" growth rates within each year at $\alpha = 0.05$. Source: adapted from Kiniry et al. [14].

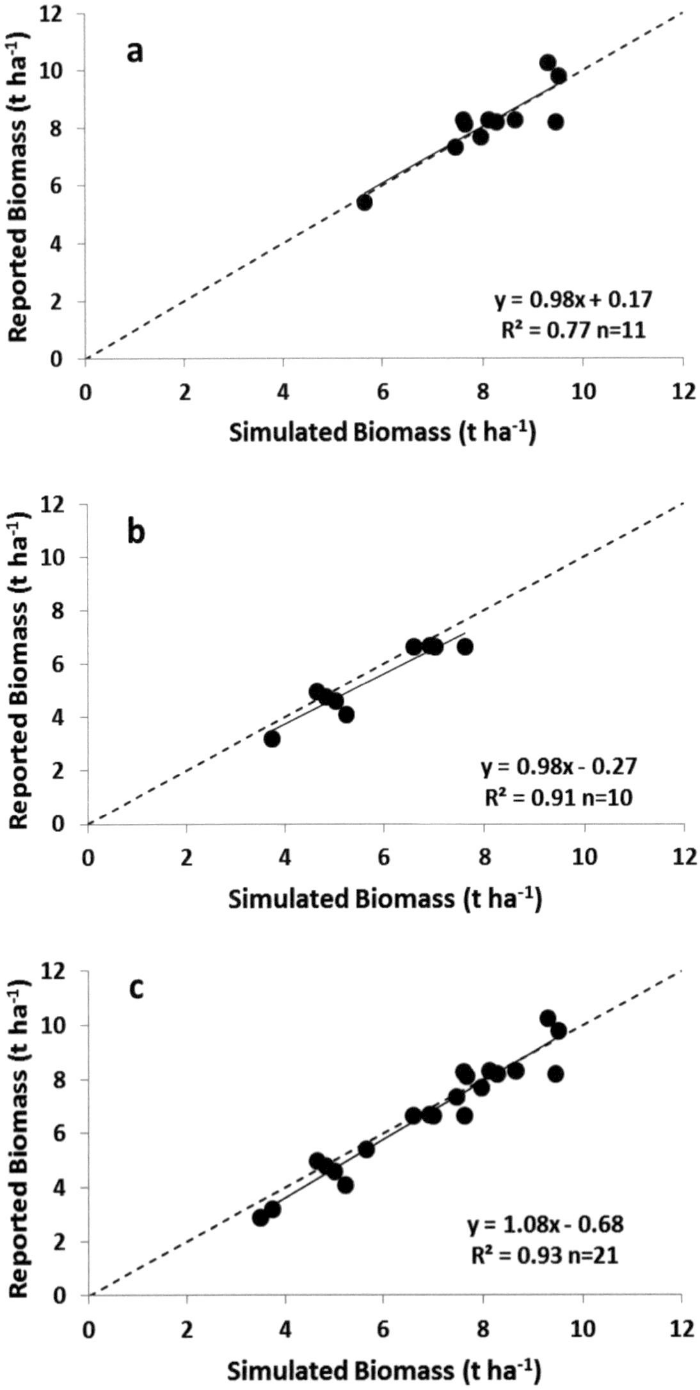

Figure 2. Reported (USDA-NRCS) and simulated tall fescue yields for (a) high-yielding soils, (b) low-yielding soils, and (c) high- and low-yielding soils at diverse sites in the U.S. Source: adapted from Kiniry et al. [14].

http://dx.doi.org/10.5772/intechopen.79987

developed to simulate competition between crops and weeds and has been applied to communities of plants such as native range sites and woody plants competing with forages. Aspects of competition simulated in ALMANAC include competition for light, water, and nutrients.

The fraction of incoming solar radiation intercepted by the leaf canopy (FI) is:

$$\mathrm{FI} = 1.0 - \exp\left(-\mathrm{k}^{*}\mathrm{LAI}\right) \tag{1}$$

The light extinction coefficient (k) for Beer's law [16] is calculated for each harvest date as:

$$\mathrm{k} = [logn\ (1\ FIPAR)]/\mathrm{LAI} \tag{2}$$

where logn = natural log of the number, and FIPAR = fraction of IPAR.

The value of k has been determined for a number of forages in the U.S. [17–20]. Realistic simulation of LAI is critical for these equations describing light interception. This is true for both the increase of LAI during active growth and the decline as leaves senesce. The model uses an S-curve to simulate the accumulation of leaf area increase as a function of GDD.

Similarly, as described above, biomass growth is simulated with a radiation use efficiency (RUE) approach [17, 21]. The RUE is calculated as the rate of increase in dry matter (g per m^2 ground area) per unit of intercepted photosynthetically active radiation (IPAR) (MJ per m^2 ground area). Regressions are fit with the treatment means of plant dry weight and summed IPAR for each sampling point. The RUE is the slope of the regression for this plant weight (g m^{-2}) as a function of the summed IPAR (MJ m^{-2}).

This regression is ideally based on multiple harvest dates during the active growth period of the forage. Occasionally, when only two harvest dates are usable, RUE is calculated from differences. Only data from dates showing increases in dry matter (actively growing) are included. This constrains RUE values to periods of active growth. Data from sites experiencing drought stress are avoided. Values for FIPAR are calculated on a daily basis, with values for dates between measurement dates calculated by linear interpolation.

Simulated light competition uses functions of [22], whereby the light interception of each plant species in the mixture is computed with the following formula: LAI**k* (*k* being the light extinction coefficient). These products (LAI*k) of each species are summed and the sum used in Beer's law to compute the fraction of light interception by the whole plant community. This fraction of light intercepted for the whole plant community is then divided among the competing species by weighted fractions. The weights account for differences in species heights and LAI**k* of the species. Thus, taller species and those with higher LAI values and higher *k* values intercept a greater fraction of the total light intercepted by the plant community.

Water and nutrient competition are simulated with a balance sheet approach. Once intercepted light for each plant species is computed as described above, the potential daily biomass growth is calculated for each species with the total daily incident solar radiation, assuming 45% of that is PAR [23, 24]. The RUE for the species multiplied by the intercepted PAR is the potential biomass growth on any given day. Using the optimum nutrient concentrations for N and P at the current growth stage, the demands for N and P are calculated. If insufficient N and/or P is present in the current rooting zone, the model reduces simulate growth rates to account for N and/or P stress. This simulation of the balance of nutrients is done for each species within a mixture. The model accounts for variability in root scavenging capacities between species only through differences in the current rooting depth of each species. Potential rooting depths of various plant species are derived from measurements reported in the literature for forages grown on soils with no restrictive soil layers (such as in [17]).

Likewise, potential plant transpiration is calculated from the potential evapotranspiration and the total community LAI. If soil water in the current rooting zone is insufficient to meet the species' demand, simulated drought stress occurs and limits growth. This occurs for all plant species present. However, it should be noted that a deeper rooted plant species may have access to soil water (and nutrients) not available to any competing shallower rooted species; ALMANAC accommodates different rooting depths of species. The deeper rooted plant species may have adequate soil water and nutrients to avoid drought and nutrient stresses when a shallower rooted species is stressed. The ALMANAC model does not currently simulate hydraulic lift dynamics and the potential impacts of lift on water and nutrient redistribution.

3. Soils and weather data

The soils and weather data described below are specific to the U.S. These are described in more detail, including how to download them at: https://www.ars.usda.gov/plains-area/temple-tx/-grassland-soil-and-water-research-laboratory/docs/193226/. The soil and weather data for the country of Mexico have also been developed and formatted for ALMANAC model simulations (https://www.ars.usda.gov/plains-area/temple-tx/grassland-soil-and-water-research-laboratory/docs/almanacmex/). As the model is applied outside of these two countries, the input data for soils and weather (as well as plant species growth curves) can be developed through cooperation with the senior author of this project.

The general philosophy of input data development is to make this model and other USDA-ARS models (including EPIC [25], APEX [26, 27], and SWAT [28–30]) readily and easily applied. Input data are constrained by what is readily available and easily accessible. This means the daily weather inputs required consist of maximum and minimum temperature, rainfall amounts (and snowfall amounts), and solar radiation. When unavailable for a given location, solar radiation can be derived; wind speed and relative humidity can be used to approximate solar radiation. Weather data from the U.S. National Oceanic and Atmospheric Administration (NOAA) websites are readily downloaded for any state in the U.S. via the steps outlined in the model documentation (see website link above).

Similarly, required soil data are available through USDA-NRCS (https://www.nrcs.usda.gov/wps/portal/nrcs/main/soils/survey/), which has the most extensive and verified, publicly-available soil database for the U.S. The soil data are readily downloaded for any state in the U.S., with the steps outlined in the ALMANAC model documentation. The most critical components of the soil data inputs are the depth, texture, and amount of rocks by soil layer.

For each soil layer, the values for saturation, drained upper limit, and lower limit are used by the model. Soil organic matter is another input that impacts plant-available water and soil carbon balances in the model. The amount of runoff from rainfall events is calculated with the traditional runoff curve number system. The runoff is simulated with the slope and type of ground cover.

4. Deriving plant parameters for a forage species and accommodating ecotypes

4.1. Field plant species measurements

The group of readily derived plant parameters includes the potential leaf area index (LAI), the development curve for LAI over the growing season, the light extinction coefficient for Beer's law (*k*), the radiation use efficiency (RUE), the duration of the season in degree days, the harvest index for seeds (HI), and the N and P concentrations for each species over the growing season. All of these should be derived from measurements of a plant stand grown in a relatively stress-free environment to establish potential values for these for each forage species and ecotype. This means that ideally species being measured in field conditions should not have stresses due to drought or nutrient deficiency.

Details on taking field measurements for deriving plant parameters are outlined in detail under the headings: "Gathering Field Data, How to Use Ceptometer: AccuPAR LP-80 Basics Standard "[31] and "Taking measurements for ALMANAC: Sampling Protocol Standard with Photos" (https://www.ars.usda.gov/plains-area/temple-tx/grassland-soil-and-water-research-laboratory/docs/193226/) (**Figure 3**).

Field-derived values for the critical species-specific parameters have been described previously [17–21]. The model simulates light interception by the leaf canopy with Beer's law [16]

Figure 3. Intercepted photosynthetically active radiation (IPAR) measurements using an AccuPAR LP-80 Ceptometer at Bishop, California, and Bryan, Texas.

and the LAI. Larger values of the extinction coefficient have more light intercepted at a given LAI.

Measurement of light interception by the plant canopy is described at the website: https://www.ars.usda.gov/plains-area/temple-tx/grassland-soil-and-water-research-laboratory/docs/193226/. To derive leaf area, biomass, and the extinction coefficient for Beer's law, LAI measurements are derived every 2 weeks during the active growing season via light measurements taken above and below the canopy between 10 a.m. and 2 p.m. on a clear day. The Decagon ceptometer (or something similar) is used to measure light as photosynthetically active radiation, since those are the wavelengths critical for photosynthesis. A random sample area for the area of interest is chosen where the forage is growing. The stand in the area for taking light measurements should not be trampled. Areas adjacent to where previous samples were taken should be avoided and should be ungrazed. A quadrat 0.5 m wide by the length of our light bar (0.8 m) is reasonable for the sampling area.

If there are any non-targeted plants in or overshadowing our quadrat, they should be removed, or the quadrat should be relocated. Only canopy cover from targeted specie should be measured. The time of day, average phenology, and the average plant height in centimeters should be recorded. Light interception readings using the ceptometer are taken as:

a. Select an area under direct sunlight near our plots, and level the external sensor on the tripod. (*Note*: Whenever moving the tripod, level the sensor and calibrate again.)

b. Calibrate the light bar with the external sensor. Take at least 10 measurements with the light bar under direct sunlight. (*Note*: Make sure measurements are taken facing the sun, thus avoiding shading the light bar or the external sensor.) Record the shown average of all 10 measurements on the datasheet.

c. When taking light measurements under the canopy using the ceptometer, take at least six evenly spaced measurements in each quadrat near ground level. Record the average. Always take care to avoid biasing the sample in favor of more plants or more bare ground.

d. Finally, harvest plants, removing all plant material in the quadrat directly above the site where light was measured and place in labeled bag.

Repeat these steps at least three more times for a targeted plant species. For each set of measurements, make sure to measure plants on the same soil or ecological site. When returning to the general area for future measurements, select the same species to measure but not the exact same plant/plot area as previously measured.

Process plant material as soon as possible after sampling to avoid desiccation effects on leaf area.

a. When weighing the entire sample from field, if the entire sample is greater than 100 g, take a representative subsample. This is between 10 and 30% of the entire sample but no less than 100 g. Weigh and record the subsample weight. Make sure to select a subsample with the same proportion of green leaves, dead material, stems, and reproductive structure as the entire sample.

b. Use a belt-driven leaf area meter (or something similar) to measure leaf area of the subsample. Separate the subsample into dead material (anything completely brown), stems, leaves, and reproductive structures. Record the weight of the dead material, stems, and reproductive structures.

c. Determine the area of each structure using the leaf area meter. Run the dead material, stems, leaves, and reproductive structures through separately and record the area of each. Place the entire sample into a paper bag. These samples are dried in a 66°C forced air oven for 3 days or until weight stabilizes. Then record the dry weight of the entire sample.

d. Finally, grind dry sample to prepare for nutrient analysis.

5. Model testing against independent data

Following successful calibration of the ALMANAC model with field measured parameters, the model was applied to simulate forage yields across years and diverse environments in the U.S. For model testing, we used published forage yield data from Natural Resources Conservation Service, United States Dept. of Agric. 2017. Web Soil Survey. Available online: http://websoilsurvey.sc.egov.usda.gov/App/WebSoilSurvey.aspx.

Many common native and introduced grasses or grass mixtures in the U.S. have annual productivity values reported as USDA-NRCS ecological site productivity (for native forages) or NRCS crop productivity (for improved grasses) for many representative areas. As discussed below, once plant parameters for a particular forage are derived, they are tested on different soils in contrasting U.S. counties. The counties simulated are selected because they have soils with quantified annual biomass yields for the forage of interest (NRCS Web Soil Survey) (http://websoilsurvey.nrcs.usda.gov/app/HomePage.htm).

Total annual production of forages reported by NRCS are derived from end-of-season sampling on sites with closed canopy stands of the species of interest over 3 years or more. The NRCS procedure involves measuring dry matter biomass production above a 5-cm cutting height in at least 10 randomly selected plots at each field site. The specific soils for a location of interest can be downloaded as described above. Mean simulated forage yield over 10 years of real weather data can be compared to the reported annual production (from USDA-NRCS Web Soil Survey) for a site. The NRCS value of Animal Unit Month (AUM) is converted to Mg ha^{-1} (0% moisture) with a conversion factor assuming 700 lbs. (318 kg) of air-dried biomass (90% moisture) per AUM. Values for key plant parameters for the plant species of interest are derived from the field measurements described above.

5.1. Examples of testing ALMANAC's simulation of forage yields

We have several published examples of testing ALMANAC's simulation of forage yields. The first was for several Texas range sites with native warm-season grasses [32, 33]. Next, we simulated old world bluestems (*Bothriochloa* Kuntze, *Capillipedium* Stapf, and *Dichanthium*

Willemet) and buffelgrass (*Pennisetum ciliare* (L.) Link) in Oklahoma, Texas, and Mexico [18]. To evaluate the ability of the model to simulate introduced or improved grasses, we tested coastal bermudagrass (*Cynodon dactylon* (L.) Pers.) and bahiagrass (*Paspalum notatum* Flügge var. *saurae* Parodi) at several sites in Texas [19]. Western grasses in low-rainfall sites in Montana were simulated using parameters derived for some common native grasses there [20]. The cool-season forage "tall fescue" was simulated at several sites in several states where this grass is commonly grown [14]. In addition, creosote bush (*Larrea tridentata* [DC.] Cov.) parameters were derived the and model testing for its ability to describe competition of this woody species with forages in arid sites in western Texas [34].

Overall, the ALMANAC model predicted forage yields with reasonable accuracy, and hence when fully calibrated, the model can be used as an effective management tool to evaluate management practices that maximize forage yields, optimize inputs, and minimize negative environmental outcomes.

6. Knowledge gaps and areas for future improvement as a guide for additional research

The ALMANAC model uses the best plant growth modeling functions currently developed. Often, knowledge gaps force model developers to use placeholder functions with the hope that future research will enable development of improved, more realistic functions. Some areas for beneficial future research include nutrient and carbohydrate cycling, forage regrowth following haying, nutrient response functions, and legacy effects.

The simulated cycling of nutrients in the soil and between the roots to the shoots for perennials needs to be critically investigated for this model. As forages mature and leaves senesce during the fall and winter, often nutrients and carbohydrates are translocated back into the root system, to be used for regrowth the following spring. Grazing may also trigger plants to allocate more carbohydrate storage in roots to survive grazing pressures. Functions describing these processes need to be better developed and incorporated into the ALMANAC model in the future.

Likewise, the regrowth of forages following hay cutting or grazing within the growing season, needs to be more extensively tested. The functions currently in ALMANAC appear to function reasonably. However, as more extensive data are available for testing the model, improvements likely will be made.

The response of forages to applied nutrients often is highly dependent on what is already in the soil. This includes nutrients readily available and those coming from transformations within the soil during the growing season. Very often publications report a nutrient response of a forage without adequately describing initial soil conditions. If adequate nutrients are already present in the soil, the response of the forage to applied nutrients can be much dampened. Likewise, if the soil is initially very nutrient poor, the forage may show a large response to applied nutrients. An extensive testing of the model with data having good values for initial soil nutrients will be valuable.

Finally, legacy effects due to previous years' weather conditions and previous years' nutrient cycling need to be investigated. This has been studied with switchgrass [35], but needs more extensive studies with diverse representative forages.

7. Conclusions

In this chapter, we described the ALMANAC model, including the process simulated, how to derive plant parameters for additional forage species, and how to validate using measured field data. Because of its accurate simulation of plant production, the water balance, and the nutrient balance, the model is capable of simulating a wide variety of environmental and management impacts on forage production, soil health, and conservation concerns, including nutrient and sediment losses. The model will be a useful and valuable tool for forage management in pastures and rangelands in a wide range of conditions.

Acknowledgements

This material is based upon work supported by the Natural Resources Conservation Service, U.S. Department of Agriculture, Conservation Effects Assessment Project for Grazing Lands, under interagency agreement number 67-3A75-13-129. This work was also supported in part by an appointment to Agricultural Research Service administered by Oak Ridge Institute for Science and Education through interagency agreement between U.S. Department of Energy (DOE) and U.S. Department of Agriculture (USDA), Agricultural Research Service Agreement #60-3098-5-002. The authors are grateful to many university, USDA-ARS, and USDA-NRCS collaborators that are listed as coauthors in the cited references by the senior author.

Conflict of interest

Authors have declared that no competing interests exist.

Author details

James R. Kiniry[1]*, Sumin Kim[2], M. Norman Meki[3] and Mari-Vaughn V. Johnson[4]

*Address all correspondence to: jim.kiniry@ars.usda.gov

1 U.S. Department of Agriculture, Agricultural Research Service, Grassland, Soil and Water Research Laboratory, Temple, TX, USA

2 Oak Ridge Institute for Science and Education, Oak Ridge, TN, USA

3 Texas A&M AgriLife Research, Temple, TX, USA

4 U.S. Department of Agriculture, Natural Resource Conservation Service, Temple, TX, USA

References

[1] Mohtar R, Buckmaster DR, Fales SL. A grazing simulation model: GRASIM, a model development. Transactions of the American Society of Agricultural Engineers. 1997;**40**:1483-1493

[2] Mohtar R, Zhai T, Chen X. A world wide web-based grazing simulation model (GRASIM). Computers and Electronics in Agriculture. 2000;**29**:243-250

[3] Wight JR, Skiles JW. SPUR, Simulation of Production and Utilization of Rangelands: Documentation and User Guide. Publication ARS-63. Washington, DC: USDA-ARS; 1987. p. 367

[4] MacNeil ND, Skiles JW, Hanson JD. Sensitivity analysis of a general rangeland model. Ecological Modelling. 1985;**29**:57-76

[5] Stout WL. Evaluation of the SPUR model for grasslands of the northeastern United States. Agronomy Journal. 1994;**86**:1001-1005

[6] Innis GS. Grassland Simulation Model. New York, NY USA: Springer-Verlag; 1978. p. 298

[7] Rotz CA, Buckmaster DR, Metens DR, Black JR. DAFOSYM: A dairy forage system model for evaluating alterantives in forage conservation. Journal of Dairy Science. 1989;**72**:3050-3063

[8] Foy JK, Teague WR, Hanson JD. Evaluation of the upgraded SPUR model (SPUR2.4). Ecological Modeling. 1999;**118**:149-165

[9] Corson MS, Skinner RH, Rotz CA. Modification of the SPUR rangeland model to simulate species composition and pasture productivity in humid temperature regions. Agricultural Systems. 2006;**87**:169-191

[10] Corson MS, Skinner H, Rotz CA. Modification of the SPUR rangeland model to simulate species composition and pasture productivity in humid temperate regions. Agricultural Systems. 2017;**87**:169-191

[11] Kiniry JR, Williams JR, Gassman PW, Debaeke P. A general, process-oriented model for two competing plant species. Transactions of the American Society of Agricultural Engineers. 1992;**35**:801-810

[12] Kallenbach RL, Bishop-Hurley GJ. A Guide to the Common Forages and Weeds of Pastures. Outreach and Extension. Columbia: University of Missouri; 2002. p. 49

[13] Kallenbach RL. Describing the dynamic: Measuring and assessing the value of plants in the pasture. Crop Science. 2015;**55**:2531-2539

[14] Kiniry JR, Kim S, Williams AS, Lock TR, Kallenbach RL. Simulating bimodal tall fescue growth with a degree-day based process-oriented plant model. Grass and Forage Science. 2018:**73**:432-439

[15] Kiniry JR, Major DJ, Izaurralde RC, Williams JR, Gassman PW, Morrison M, Bergentine R, Zentner RP. EPIC model parameters for cereal, oilseed, and forage crops in the northern Great Plains region. Canadian Journal of Plant Science. 1995;**75**:679-688

[16] Monsi M, Saeki T. Uber den lichtfaktor in den pflanzengesellschaften und seine bedeutung fur die stoffproduktion. Japanese Journal of Botany. 1953;**14**:22-52

[17] Kiniry JR, Tischler CR, Van Esbroeck GA. Radiation use efficiency and leaf CO_2 exchange for diverse C_4 grasses. Biomass and Bioenergy. 1999;**17**:95-112

[18] Kiniry JR, Johnson MV, Venuto BC, Burson BL. Novel application of ALMANAC: Modelling a functional group, exotic warm-season perennial grasses. American Journal of Experimental Agriculture. 2013;**3**:631-650

[19] Kiniry JR, Burson BL, Evers GW, Williams JR, Sanchez H, Wade C, Featherston JW, Greenwade J. Coastal bermudagrass, bahiagrass, and native range simulation at diverse sites in Texas. Agronomy Journal. 2007;**99**:450-461

[20] Kiniry JR, Muscha JM, Petersen MK, Kilian RW, Metz LJ. Short duration, perennial grasses in low rainfall sites in Montana: Deriving growth parameters and simulating with a process-based model. American Journal of Experimental Agriculture. 2017;**15**:1-13

[21] Kiniry JR, Evers GW. Radiation use efficiency of arrowleaf, crimson, rose, and subterranean clovers. Agronomy Journal. 2008;**100**:1155-1160

[22] Spitters CJT, Aerts R. Simulation of competition for light and water in crop–weed associations. Aspects of Applied Biology. 1983;**4**:467-483

[23] Monteith JL. Light distribution and photosynthesis in field crops. Annals of Botany (London). 1965;**29**:17-37

[24] Meek DW, Hatfield JL, Howell TA, Idso SB, Reginato FJ. A generalized relationship between photosynthetically active radiation and solar radiation. Agronomy Journal. 1984;**76**:939-945

[25] Williams JR, Jones CA, Dyke PT. A modeling approach to determining the relationship between erosion and soil productivity. Transactions of ASAE. 1984;**27**:129-144

[26] Williams JR, Potter S, Wang X, Atwood J, Norfleet L, Gerik T, Lemunyon J, King A, Steglich E, Wang C, Pitts T, Meinardus A. APEX model validation for CEAP. 2010. Available from: http://www.nrcs.usda.gov/technical/nri/ceap/umrbdocumentation/

[27] Gassman PW, Williams JR, Wang X, Saleh A, Osei E, Hauck LM, Izaurralde RC, Flowers JD. The Agricultural Policy Environmental Extender (APEX) Model: An emerging tool for landscape and watershed environmental analyses. Transactions of the American Society of Agricultural and Biological Engineers. 2010;**53**:711-740

[28] Arnold JG, Srinivasan R, Muttiah RS, Williams JR. Large area hydrologic modeling and assessment, Part I. Model development. Journal of the American Water Resources Association. 1998;**34**:73-89

[29] Arnold JG, Fohrer N. SWAT2000: Current capabilities and research opportunities in applied watershed modeling. Hydrological Processes. 2005;**19**(3):563-572

[30] Gassman PW, Reyes MR, Green CH, Arnold JG. The soil and water assessment tool: Historical development, applications, and future research directions. Transactions of the American Society of Agricultural and Biological Engineers. 2007;**504**:1211-1250

[31] Johnson MV, Kiniry JR, Burson BL. Ceptometer deployment method affects measurement of fraction of intercepted photosynthetically active radiation. Agronomy Journal. 2010; **102**:1132-1137

[32] Kiniry JR, Sanderson MA, Williams JR, Tischler CR, Hussey MA, Ocumpaugh WR, Read JC, van Esbroeck GV, Reed RL. Simulating Alamo switchgrass with the ALMANAC model. Agronomy Journal. 1996;**88**:602-606

[33] Kiniry JR, Sanchez H, Greenwade J, Seidensticker E, Bell JR, Pringle F, Peacock G, Rives J. Simulating grass productivity on diverse range sites in Texas. Journal of Soil and Water Conservation. 2002;**57**:144-150

[34] Kim S, Kiniry JR, Loomis L. Creosote bush, an arid zone survivor in Southwestern U.S.: 1. Identification of morphological and environmental factors that affect its growth and development. Journal of Agriculture and Ecology Research International. 2017;**11**:1-14

[35] Hawkes CV, Kiniry JR. Legacies in switchgrass resistance to and recovery from drought suggest that good years can sustain plants through bad years. Bioenergy Research. 2018; **11**:86-94

Chapter 4

Mob vs. Rotational Grazing: Impact on Forage Use and *Artemisia absinthium*

Heidi Reed, Alexander Smart, David E. Clay,
Michelle Ohrtman and Sharon A. Clay

Additional information is available at the end of the chapter

http://dx.doi.org/10.5772/intechopen.79085

Abstract

Short duration (≤24 h), high stocking density grazing systems (e.g., mob grazing) mimics historic prairie grazing patterns of American bison (*Bison bison*), and should minimize selective grazing. We compared mob [125 cow-calf pairs on either 0.65 ha for 12 h; or 1.3 ha for 24 h] vs. rotational [25 cow-calf pairs on 8.1 ha for 20 days starting in mid-May with or without 2,4-D application prior to grazing; and 15 days starting mid-April (no herbicide)] grazing systems based on forage utilization and impact to *Artemisia absinthium* (absinth wormwood) in a tall grass pasture of Eastern South Dakota. Grass height and density, and *Artemisia absinthium* patch volume were quantified pre- and post-grazing at sampling points along multiple transects. Mob grazing had >75% forage utilization, whereas rotational grazing averaged 50% (all consumption). Within a grazing season, three grazing systems suppressed *Artemisia absinthium* patches with rotation/spray (100% decrease) > mob (65 ± 10% decrease) > mid-May rotation (41 ± 16% decrease), whereas *Artemisia absinthium* patches in the mid-April rotation followed by summer rest dramatically increased in size. *Artemisia absinthium* patches <19,000 cm^3 were browsed, whereas larger patches were trampled in mob-grazed areas, but avoided in rotational grazing. All *Artemisia absinthium* patches had regrowth the year following any grazing event.

Keywords: cattle, forage utilization, mob grazing, rotational grazing, weed management

1. Introduction

Grazing lands are managed to optimize forage and animal productivity, and minimize adverse impacts to soil and the surrounding environment. The annual economic impact of all weedy species in U.S. grazing lands is greater than all other pests combined [1], and has been estimated at 1 billion dollars for forage loss and 5 billion dollars for control costs [2]. Weed infestations cause a variety of problems in grazing lands. Weeds can reduce forage vegetative quality and quantity; displace native plants and animals; reduce animal fertility, weight gains, or be toxic, resulting in fatalities; reduce meat and/or hide quality; increase management costs; and reduce land values [3, 4]. Tactics for weed management in pastures and grazing lands vary with the type of weed, livestock species, and applicability of other methods (e.g., mowing, biocontrol, herbicide treatment) [5, 6].

Livestock can help manage weeds by grazing or trampling and can improve pasture condition and competitiveness of desirable plants by increasing soil nutrients through manure and urine deposition [3]. Weed species and stage of growth; livestock species; and stocking rate and duration influence grazing effectiveness on weeds [3, 7]. Unfortunately, cattle (*Bos taurus*), the grazing livestock of choice in the Northern Great Plains (NGP), selectively consume forage in dung-free areas [8, 9], and avoid weeds for a wide variety of reasons [10]. Cattle herds are not managed specifically for weed control for several reasons. First, cattle are expensive to raise and replace and, even with premium prices, the economic margin is narrow [7]. Weeds may not be as palatable as grasses, and lower consumption may reduce weight gains [7], or, if high in alkaloids, problems with reproduction and/or toxicity can occur [11].

Rotational grazing often uses a 'take half'- 'leave half' forage philosophy to maintain healthy, vigorous plant communities [12, 13]. Mob grazing has been promoted to mimic the world's historic grassland ecosystems [14] with herds of large animals intensively grazing areas and moving often. The definition of mob grazing is subjective, but typically includes using extremely high stocking rates (100 head or more per ha) for short periods of time (moving every 12 or 24 h) [15] followed by recovery periods of 6–12 months. The goal of mob grazing is to have every plant within the enclosure eaten [16] or trampled [17], limiting selectivity or avoidance of specific species [9], and providing a more homogeneous grazing treatment. Barnes et al. [16] reported that grazing homogeneity correlated with paddock size, with pastures ranging from 1 to 8 ha in size grazed nearly uniformly, even if the same stocking rate per ha are used on larger areas.

Grazing impact for weed management is maximized when the target weed is most palatable, is the only forage option, or is made more palatable to livestock in some way (e.g., salt or sugar treatment) [7], and the desired vegetation is at its least vulnerable phenotypic stage [1]. High animal densities maximize trampling, which incorporates plant litter, manure, and urine into soil, increasing organic carbon and soil nutrients [17]. The combination of eating, trampling, and long rest periods is expected to increase productivity of more desirable forage [3, 18].

Mob grazing has been adopted by ranchers in Texas, SE Colorado, central Nebraska, Missouri, and other areas [19] where vegetative regrowth can occur quickly due to warm conditions, and high rainfall or irrigation capabilities. Under dryland conditions of the NGP, timing mob grazing to fit within the vegetative and environmental constraints of the area is difficult as growing seasons are short, and pastures often experience summer drought. McCartney and Bittman [20] reported on a mob grazing study that used 7–14 heifers ha^{-1} (dependent on seasonal timing) on about 0.3 ha paddocks at different intensities (light, grazed twice a year; to intense, grazed five times a year) in northeastern Saskatchewan. They observed positive [decline in smooth brome (*Bromus inermis*)], negative [decline of intermediate wheatgrass (*Elytrigia repens*), and increase bluegrass (*Poa* sp.) species], or no effects on specific species [e.g., green needlegrass (*Nassella viridula*)] over 4 years. These findings suggest that intensive grazing benefits are related to plant species, stocking density, and grazing timing, all of which can be manipulated for maximum impact [21]. Ranchers interested in using mob grazing for increased productivity and harvest efficiency would benefit from on-farm research that examines the relationship among stocking densities and timing on vegetative utilization and the impact to locally invasive weed species.

There have been few comparisons in the NGP among mob grazing and other, more conventional, grazing systems. Fundamental problems in grazing research often include small enclosure sizes and animal numbers, which provide data that are difficult to scale to commercial operations [16]. Due to the expense, need for many animals, and labor and time involved to move cattle frequently, the study was managed by an Eastern South Dakota rancher who incorporates both rotational and mob grazing techniques into his cattle operation. *Artemisia absinthium* was selected as a model invasive plant as it is a non-native perennial forb that cattle typically avoid due to woody stems of older plants, and unpalatability due to the production of secondary compounds and essential oils [22]. The objectives of this study were to (1) quantify forage present before grazing (pre-graze) to be able to estimate forage utilization post-graze in mob and rotational grazing systems, (2) determine the impact of grazing system on *Artemisia absinthium* suppression in-season based on mob and rotational grazing and (3) examine the recovery of *Artemisia absinthium* patches a year after mob grazing.

2. Grazing impacts to forage utilization and Artemisia absinthium

2.1. Experimental site

The effects of rotational and mob grazing stocking densities on Artemisia absinthium and surrounding forage utilization were compared in an eastern South Dakota rangeland location in the tall grass prairie habitat near Hayti (44.66°N, −99.22°W) in 2013 and 2014 [23]. The dominant soil series of the rotationally grazed pasture were the: Poinsett-Waubay silty clay loams (Calcic Hapludolls/Pachic Hapludolls); Buse-Poinsett complex (Typic Calciudolls/Calcic Hapludolls); and Poinsett-Buse-Waubay complex (Calcic Hapludolls/Typic Calciudolls/Pachic Hapludolls) [https://soilseries.sc.egov.usda.gov/osdname.aspx]. Mob grazing pasture soils were similar to the rotational pasture with the addition Barnes-Buse loam complex (Calcic Hapludolls/Typic Calciudolls). The plant communities in these pastures were a mix of cool season native and invasive grasses, warm season grasses, and broadleaf species (**Table 1**).

2.2. Weather

Growing degree days (GDD) were calculated to provide a reference for plant development between sampling dates and years. The GDD calculation [GDD = $\sum$ (maximum daily temperature + minimum daily temperature)/2 − base temperature] used the base temperature of 0°C, due to majority of cool season species with GDD accumulations starting on January 1 of each year.

Mob-grazed sites		Rotational sites	
Common name	Scientific name	Common name	Scientific name
Big bluestem	*Andropogon gerardii*	Western wheatgrass	*Pascopyrum smithii*
Sweet clover	*Melilotus officinalis*	Absinth wormwood	*Artemisia absinthium*
Alfalfa	*Medicago sativa*	Smooth brome	*Bromus inermis*
Red clover	*Trifolium pratense*	Kentucky bluegrass	*Poa pratensis*
Kentucky bluegrass	*Poa pratensis*		
Dandelion	*Taraxacum officinale*		
Absinth wormwood	*Artemisia absinthium*		
Western wheatgrass	*Pascopyrum smithii*		
Smooth brome	*Bromus inermis*		

Table 1. Plant species in the mob-grazed and rotational grazed sites at Hayti, SD in 2013 and 2014.

Precipitation (from January 1) was also determined. The rotational pre-graze samples in 2013 were taken on June 13, with 641 GDD and 243 mm precipitation (www.noaa.gov) with values similar to the 30-year (1980–2010) average. Post-grazing samples were taken July 22, with GDD of 1540 and precipitation totaled 343 mm. In 2014, samples were taken May 13 with GDD at the spring assessment (which was taken after the early spring grazing) 262 and 65.5 mm of precipitation. The fall assessment was taken September 16 with GDD of 2603, and total rainfall of 370 mm fall. Rotational grazing was done much earlier in 2014 because the rancher was concerned about low amounts of precipitation (nearly 60% below average) during the 2013 fall and winter.

GDD accumulations for mob-grazed areas in 2013 were 1801 (August 6) and 1855 (August 9) for pre- and post-graze samples, respectively. Precipitation totaled 376 mm before and after mob grazing. In 2014, GDDs were 1693 pre-graze (July 29) and 1817 post-graze (August 4) and precipitation for pre-graze and post-graze totaled 230 and 270 mm, respectively.

2.3. Grazing treatments

Stocking treatments (rotation vs. mob) were repeated, although cattle densities and time of grazing differed between the 2 years due to feeding needs and differences in forage growth due to low rainfall in 2014 (**Table 2**). Rotational grazing was conducted in 8 ha pastures with 25 cow-calf pairs (1560 kg ha^{-1}). In 2013, in one paddock, the cow-calf pairs were allowed to graze for 14 days starting June 13 (referred to as 'rotation'). In a separate paddock, generic 2,4-D ester at 1.1 kg ha^{-1} [24] was applied 1 day before the start of grazing on June 13 with a grazing duration of 14 days (referred to as 'spray/rotation'). In 2014, a different pasture was grazed by 25 cow/calf pairs for 15 days, starting April 27 and ending May 11 (referred to as 'early spring grazed/summer rest').

Mob grazing was conducted for 12 h in a 0.65-ha paddock on August 8, using 125 cow-calf pairs (stocking rate of 223,250 kg ha^{-1} day^{-1}) (**Figure 1**). In 2014, a different 1.3-ha area was mob grazed on July 30 for 24 h with 125 cow-calf pairs (stocking rate of 53,580 kg ha^{-1} day^{-1}).

http://dx.doi.org/10.5772/intechopen.79085

				Pre-graze		Post-graze				
							Biomass[1]			
Year	Grazing system	Stocking density	Grazing duration	Sampling date	Forage biomass[1]	Sampling date	Standing	Trampled[2]	Forage use efficiency[3]	Forage utilization[4]
		kg ha^{-1}			kg ha^{-1}		kg ha^{-1}		%	%
2013	Mob	223,250	12 h	6-Aug	2910^{a5}	9-Aug	570^{b}	530^{b}	62	80
	Rotation	1560	20 days	13-Jun	2600^{a}	22-Jul	1190^{b}	0	45	45
	Rotation/spray	1560	20 days	13-Jun	4530^{a}	22-Jul	2528^{b}	0	57	57
2014	Mob	53,580	24 h	29-Jul	4640^{a}	4-Aug	1170^{b}	1890^{b}	34	75
	Rotation/ Summer ungrazed	1560	15 days		~1700[6] (estimated)	13-May 16-Sep	870 2090[7]			

[1]Vegetation biomass was estimated using the cool season mixed pasture grazing stick method, (vegetation height cm – 7.6 cm) * 79 to estimate kilograms per ha.

[2]Trampled vegetation was any plant with a stem less than a 450 angle from the soil surface.

[3]Forage efficiency (consumption only) was calculated by: [(before grazing vegetation –[standing + trampled])/(before grazing vegetation)]*100.

[4]Forage utilization (consumption + trampling) was calculated by: [(before grazing vegetation – standing vegetation after grazing)/before grazing veg]*100.

[5]Values with different letters within the same row for the pre graze vegetative biomass compared with post-graze standing or trampled (mob) or total vegetative biomass (rotational) differed at $P < 0.0001$.

[6]Samples were not taken pre-graze in this treatment but estimated from the leave half/take half grazing system.

[7]Forage in autumn following the grazing treatment in the spring.

Table 2. Mob and rotational grazing stocking density, grazing duration, sampling dates, forage biomass pre- and post-graze, and forage efficiency and utilization by year.

Figure 1. A mob grazing herd waiting for the next pasture.

2.4. Forage amounts and utilization

Eight 50-m long transects were established in each paddock for vegetative production evaluation. Sampling points were placed every 5 m along each transect, with GPS coordinates (Garmin etrex 20, Garmin, LTD, Schaffhausen, Switzerland) recorded so that resampling occurred at the same points pregrazing and post-grazing. At the sampling points, pre-graze measurements (in 2013, rotational graze and spray/rotational graze—13 June; mob graze—6 August; 2014, mob graze—29 July) included vegetation height using a grazing stick [25], and ocular estimates of basal cover of living vegetation, litter cover, and bare ground (0–100%) in a 1 m^2 area around the point. In 2013, vegetation in a 0.25 m^2 area was clipped to within 1 cm of the soil surface, and bagged ($n = 30$). Litter under the vegetation also was collected. Samples were weighed, dried at 38 C to constant weight, and dry weight of vegetative biomass and litter per unit area were calculated. The biomass values and grazing stick estimates were compared at each sampled point.

A few days after grazing (in 2013, rotational graze and spray rotational/graze—22 July; mob graze—9 August; in 2014, mob graze—4 August), the same transects and sampling points were reestablished for post-grazing measurements. Vegetation height was measured using the grazing stick, and percent trampled vegetation (e.g., new litter; defined as living vegetation oriented less than 45° from the soil surface) was estimated in the same areas as pre-graze sampling.

In 2014 due to the producer's needs, cattle grazed the designated rotational pasture in April and then this pasture was untouched for the remainder of the season (summer rest). Unfortunately, due to the early timing of the grazing in the second year, no pre-grazing measurements were taken for this pasture. Measurements occurred on 13 May, after the early season grazing was completed, and then resampled on 16 September (designated as regrowth after early spring grazed/summer rest). In addition, the transects which were sampled in 2013 were reestablished and vegetative height was quantified in May 2014 to examine recovery after grazing.

2.5. *Artemisia absinthium* measurements

Another three 50-m transects were established in each pasture with vegetative height measured pre- and post-graze every 2.5 m along the transects. *Artemisia absinthium* patches (individual plants if small or a patch if large) were selected and tagged near the base of the plant/patch every 5 m along these transect lines in each treatment (in 2013 rotation; spray/rotation; and mob graze; and 2014 rotation/summer rest and mob graze). Pre- and post-grazing grass height and *Artemisia absinthium* patch volume (height and two perpendicular widths) were measured at the same time as forage measurements in 2013. In late May of 2014, *Artemisia absinthium* patches measured in 2013 experimental pastures were inspected for recovery and shoot regrowth. The rotation/summer rest had *Artemisia absinthium* measurements taken in May, 2014 just after grazing, and again in September (as summer rest measurement).

2.6. Statistical analysis

Data analyses were performed using JMP®, Version 5.0.1, (SAS Institute Inc.). Forage amounts pre-graze were based on clipped biomass measurements and compared with the grazing stick method. The grazing stick equation, based on plant height, was:

$$\text{Estimated biomass } (\text{kg ha}^{-1}) = [\text{plant height (cm)} - 7.6 \text{ cm}] \times 79 \tag{1}$$

This estimated biomass for a cool season mixed grass pasture [26, 27]. The 7.6 cm value accounts for basal stems and leaves that would not be eaten by grazing animals. Two-tail, two-sample homoscedastic t-tests were used to compare forage biomass with the grazing stick estimates. Grazing stick estimates were found to be statistically similar to the clipping method.

Forage biomass and *Artemisia absinthium* volume were compared pre- and post-grazing and forage utilization (consumption plus trampling) was determined by examining new litter and remaining biomass at each transect point. These data were analyzed using one-tailed (post-graze < pre-graze) matched pair t-tests. Due to timing and treatment differences among rotational treatments, data were analyzed by treatment and year. Treatment differences are reported at a significance level of $P \leq 0.10$.

Binomial analysis of *Artemisia absinthium* patch volume data (yes = less volume post grazing; no = same or greater volume) using the equation:

$$[p \pm t_{(0.1)} \text{ sqrt} (p (1 - p)/n)] \tag{2}$$

was used to examine the influence of each treatment on *Artemisia absinthium* patches [28]. In the mob grazing treatments, *Artemisia absinthium* data were combined across years. To better understand the relationship between weed patch size and grazing system impact, *Artemisia absinthium* patches were separated into two volume classes (<19,000 cm^3 and >19,000 cm^3). In Myer [23], four volume classes originally were designated, but were combined into the two volumes due to similarity of results within smaller and larger size classes.

3. Measured impacts of grazing systems

3.1. Forage utilization

3.1.1. Mob grazing

Pre-graze forage coverage averaged 85% (grass and forb) in 2013 and neared 100% in 2014. In 2013, pre-graze forage biomass was estimated to be 2910 (±816) kg ha^{-1} with the clipped method and 2720 kg ha^{-1} with the grazing stick. These measurements were statistically similar. Pre-grazing biomass in 2014 averaged 4640 kg ha^{-1}, the grazing stick method estimated 3980 kg ha^{-1}, with estimates statistically similar. The discrepancy between direct biomass sampling and grazing stick can be partially explained by sampling method, as forage was cut to within 1 cm of soil level, but the grazing stick calculation subtracts 7.6 cm from forage height to account for unconsumed stubble. Whereas the clipping method provided excellent data, the process was labor intensive, slow, and required preweighing, drying, and postweighing. In addition, it was found that after mob grazing there was no biomass to clip. The grazing stick method provided a reasonable estimate of available forage.

In 2013, mob grazing forage utilization was about 80% (**Table 2**; **Figure 2**) with a harvest efficiency (amount consumed) of 62% (~1800 kg ha^{-1}). The remaining 20% of the vegetation was trampled. In 2014, the same stocking rate (125 cow-calf pairs) was used, but the area was two times larger, had about 1.5 times greater pre-graze biomass, and grazing time was doubled from 12 to 24 h. Forage utilization in 2014 was 75%, similar to 2013. The amount consumed was 1600 kg ha^{-1}, similar to the amount consumed in 2013, but due to the greater starting biomass, the harvest efficiency (percent consumed) was 34%, and the trampled amount was 40%.

3.1.2. Rotational grazing

In 2013, pre-graze forage amount averaged 2600 kg ha^{-1} and post-graze was 1190 kg ha^{-1} (**Table 2**). Both harvest efficiency (amount consumed) and utilization (amount consumed + trampled) were 45%, as new trampled litter was not observed. In the rotational/spray treatment,

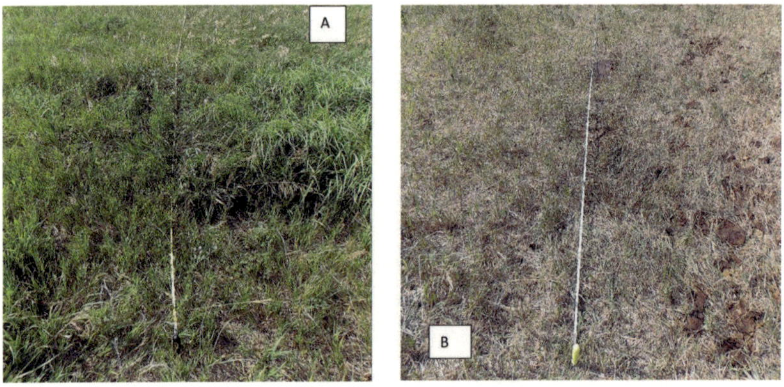

Figure 2. Pre-graze forage and post-graze results, the impact of mob grazing.

pre- and post-graze forage was 4530 and 2610 kg ha^{-1}, respectively, which indicated that forage consumption neared 57%. As in the rotational area, there was little newly trampled litter.

The 2014 rotational pasture was grazed in April, which allowed recovery during the summer/fall of 2014. Forage after grazing was 870 kg ha^{-1}. The rancher follows the 'take half, leave half' utilization recommendation [12, 13], so a reasonable pre-graze forage estimate would have been about 1300 kg ha^{-1}. Grass forage increased from 11 (May) to 23 (September) cm in height ($P < 0.001$) with fall forage biomass estimated at 2090 kg ha^{-1}.

3.2. Grazing impact on *Artemisia absinthium*

A pre-grazing assessment of *Artemisia absinthium* was conducted with the volume of the patch related to its dry biomass by recording patch volume and comparing with clipped dried biomass. In mid-June of 2013, 30 *Artemisia absinthium* patches were quantified for volume, plants clipped, and dry biomass determined before grazing. Regression analysis of biomass (expressed as log biomass +1) on plant volume (expressed as log plant volume + 1) for these 30 patches resulted in the equation: log (biomass+1) = 1.35 log (volume + 1) − 5.89, [23] which implies a direct increase in biomass as patch volume increased. This regression fit the data very well ($r^2 = 0.90$; $P < 0.001$), and was intended to be used to express differences in *Artemisia absinthium* biomass pre- and post-grazing. However, trampling dramatically increased *Artemisia absinthium* plant volume, as the shoots spread apart, in the mob-grazed areas (**Figure 3**) but because the samplings were within a few days of each other, it would not have been possible to increase biomass as the equation suggests. Therefore, data are presented and discussed in terms of plant volume, rather than biomass.

3.2.1. Mob grazing

Matched-pair analysis of 2013 and 2014 combined indicated that about 65% of the *Artemisia absinthium* patches had less volume after mob grazing (**Table 3**). In 2013 the decrease averaged 75%, whereas in 2014, the decrease was about 20%. In 2014, grass surrounding the *Artemisia absinthium* patches had 60% of the forage consumed. Therefore, it appeared that cattle were grazing close to,

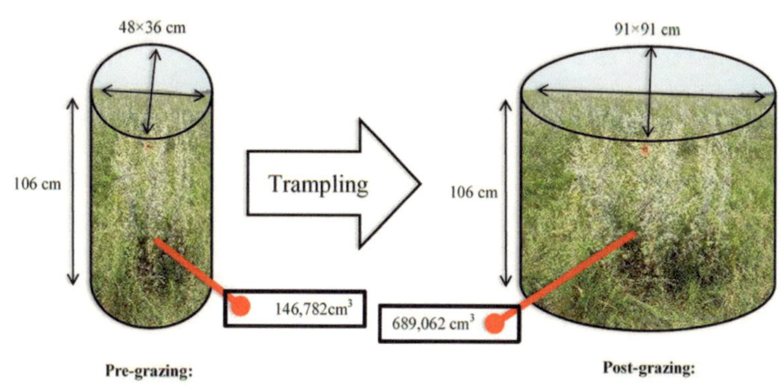

Figure 3. Example of the impact of mob grazing on a single *Artemisia absinthium* plant (pre and post grazing).

		Post-graze			
Year	Grazing system	Pre graze ave. vol.	#Decrease/total	Ave. vol. of remaining	% Control[a]
		cm³		cm³	
2013/2014	Mob	66,500	39/60	25,650	65 (10)
		<19,000	28/38		73 (9)
		>19,000	11/22		50 (NS)
2013	Rotation	12,380	12/29	8830	41 (16)
	Spray/rotation	16,500	26/27	0	100 (1)
2014	Summer recovery	2850	1/28		3 (NS)

The average pre-graze volume, the number of patches from the initial number that decreased in volume post-graze, and the average volume of the patch remaining. Patches in the mob-grazed pastures were separated into those with an initial volume < or >19,000 cm^3 and number that decreased in volume are presented.[a]Numbers in parentheses are confidence intervals based on binomial testing of the number of patches that showed a decrease over the total number with t = 0.1.

Table 3. Effect of grazing system on *Atremesia absinthium* average patch volume.

if not directly on, the *Artemisia absinthium* plants. The remaining patches increased in volume by 120% in 2013 and 154% in 2014. This volume increase at first does not seem correct, as pre- and post-grazing samples were taken within days of each other each year. However, the volume increase was due to an increase in patch width (**Figure 3**), and was attributed to trampling.

3.2.2. Rotational grazing

In 2013, 41% of the *Artemisia absinthium* patches in the rotational paddocks had a 30% decrease in volume and the remaining patches had similar volume pre- and post-grazing. Post-graze forage height of plant near the *Artemisia absinthium* patch averaged 15 cm (33%) shorter than pregrazing measurements ($P < 0.001$), which indicates that *Artemisia absinthium* may have been consumed. In the spray/rotation 2013 pasture, nearly 100% of the *Artemisia absinthium* patches decreased in volume by 100% after grazing (**Table 3**). Grass surrounding the *Artemisia absinthium* patches was 51% shorter ($P < 0.001$) post- grazing, which strongly suggests that plants in the sprayed patches were consumed with forage.

In 2014, with no grazing pressure during the summer season, only 1 (3%) of the *Artemisia absinthium* patches decreased in volume. The remainder had a volume increase of 5000% from May (average volume = 2850 cm^3) to September (average volume = 151,200 cm^3). In addition, the average height increased from 15 (May) to 86 cm (September). Because there was no trampling and an increase in shoot height, this increase can be attributed to plant growth.

3.2.3. Influence of initial Artemisia absinthium patch volume on grazing system impact

Initial *Artemisia absinthium* patch volume in the rotation and rotation/spray areas did not influence final volume. All *Artemisia absinthium* size categories in the rotationally grazed areas had about 50% of the patches increase and 50% decrease in volume. All *Artemisia absinthium* patches in the rotation/spray treatment were reduced to near 0, irrespective of initial plant volume. Initial *Artemisia absinthium* patch volume in mob-grazed areas influenced final *Artemisia absinthium* volume.

When data were combined over both years, 28 of 38 *Artemisia absinthium* patches <19,000 cm^3 decreased in volume with reductions ranging from 43 to 84% (data not shown). The other 10 patches in this category increased in volume by about 50%. There were 22 *Artemisia absinthium* patches >19,000 cm^3. Of these, 11 patches had a slight decrease in volume. The volume increase in the other 11 patches averaged 150%. Based on the height of the surrounding forage and *Artemisia absinthium* plant condition, it appears that *Artemisia absinthium* patches <19,000 cm^3 were consumed with forage, whereas plants in the larger patches were trampled and not browsed.

Artemisia absinthium plant height also was used to evaluate treatment effects. The average height of 30 *Artemisia absinthium* plants was similar before (average height 39 cm) and after (average height 37 cm) rotational grazing in 2013, which may be considered avoidance. The initial height [tall (>33 cm) vs. short (<33 cm)] did not influence rotational grazing impact. Before grazing, *Artemisia absinthium* plant height in the mob-grazed treatment averaged 58 cm. After mob grazing, 75% (± 9) of the *Artemisia absinthium* plants were 37% shorter, with no plants increasing in height. Even plants that were very tall (>97 cm) were reduced in height by about 50%. These data were consistent with either trampling or consuming. We concluded that animals in the rotation pasture had enough area and forage to selectively avoid *Artemisia absinthium* plants. Spraying 2, 4-D followed by rotational grazing (spray/rotation treatment), however, resulted in a height reduction of 96% of the *Artemisia absinthium* plants (from 54 to 9 cm).

In 2014, all tagged patches in the 2013 pastures were reevaluated to determine if patches and plants in the patches were still present and the amount of regrowth. Plants in the treated patches of the rotation/spray treatment, which provided excellent control of *Artemisia absinthium* in 2013, had less volume than those originally measured in 2013, but *Artemisia absinthium* plants were still present at the same location as the original patches (data not shown). Rotational grazing, with a 2,4-D application just prior to grazing, helped manage *Artemisia absinthium* plants in the same growing season as the herbicide application as they were no longer visible just after grazing. However, this treatment did not eliminate this perennial weed, as plants regrew the year after this treatment. Plants in the mob-grazed and rotational grazed areas were also present and had no observed injury.

4. Discussion

Rotational grazing for 20 days at 25 cow/calf pairs in 8 ha had comparable results in forage consumption to mob grazing with 125 cow/calf pairs for 12 or 24 h. in 0.65 or 1.3 ha, respectively. There were other differences between the systems, most notably the vegetative growth stage of forage, which was more mature during mob grazing. Trampled vegetation was observed in the mob grazing areas but not the rotational grazing treatments. However, claims about building soil at rates of cm per year, or significantly increasing N and C content (which was measured and reported in Myer [23]), as often discussed in popular press articles [15, 19], could not be substantiated in this study. However, trampled litter and manure patches (measured as manure patches along the transects and reported in Myer [23]) were greater post-mob grazing compared to both pre-mob and post-rotational grazing.

McCartney and Bittman [20] and others [29–31], suggest that timing and grazing capacity for optimal forage utilization and weed control, with minimal harm to desired species, requires thoughtful management to improve or maintain rangeland health. Our results show that mob grazing (225,000 or 50,000 kg of cattle ha^{-1} day^{-1}) could reduce biomass of *Artemisia absinthium* a less palatable species in a pasture. In mob-grazed treatments, *Artemisia absinthium* plants appeared to be consumed if plants were small and, most likely, still had herbaceous, rather than woody, stems. Mob grazing offers the additional benefit of trampling which reduced *Artemisia absinthium* height, although not necessarily the volume, especially of larger plants. Effectiveness of mob grazing is dependent on plants species present, stocking density, and timing [14, 16, 20]. Grazing weeds should be avoided after seed set to minimize seed dispersal, as some weed seeds remain viable or increase in germination after ingestion and passing through the digestive tract of livestock [32, 33]. While we did not find literature that specifically addresses changes in *Artemisia absinthium* seed viability after animal ingestion, *Artemisia absinthium* seeds mature in late August or September [34], after the grazing events of our study, and was not investigated. If grazing an infested pasture must be delayed until a species is past its most palatable stage, or if a weed has inherently low palatability, higher stocking rates, as seen in this study and other studies [7] improved suppression.

Mob grazing with cattle has been proposed as a grazing system to increase forage use efficiency and help in landscape restoration [14] and is likened to grazing patterns of the native plains bison. Kohl et al. [35] reported that bison and cattle differ in grazing, standing, bedding, and moving behaviors, with bison moving from 50 to 99% faster and foraging up to double the land area than cattle during the same duration. This is the precedent for the frequent moves when mob grazing cattle. In addition, cattle, when not pressured, tend to select high plant biomass, whereas bison tend to select intermediate plant biomass [35]. Regardless of the inherent differences between these two species, when managed correctly, mob grazing with cattle can diversify grazing time, with frequent moves, and long rest periods [30]. However, if managed incorrectly, high intensity grazing systems could increase weed infestations [31]. For example, in 3 years, under medium grazing intensity (grazed five times $year^{-1}$ with 6 cm of vegetation remaining after each grazing event) weeds increased by about 4 plants m^{-2}, whereas under high intensity (grazed seven times $year^{-1}$ until surface exposure), weed densities increased by 51 plants m^{-2} [36]. Hart et al. [37] reported that stocking rates that alter grazing frequency and defoliation intensity, rather than grazing system, have greater potential to impact species composition. Plant diversity and complex mixtures of forage species are integral to healthy ecosystems and consistent yields [38, 39]. However, mob grazing, if repeatedly used in the same area and at the same seasonal timing, could decrease plant species diversity and richness, change functional plant traits (e.g., tall vs. short), but improve productivity of the remaining plants [40].

The animal of choice for grazing also can influence grazing results. Goats (*Capra aegagrus hircus*) and sheep (*Ovis aries*) [7, 41] are often suggested to control brush and other undesirable vegetation, as they are more efficient at foraging and have faster growth rate than cattle. However, there are to numerous disadvantages to using goats and sheep which include: poor return on investment due to low per capita consumption of their meat products in the US and low wool prices; limited genetic improvement in milk or meat production; high predation rates compared with cattle; difficulty in fencing confinement; and susceptibility to internal parasites, which discourages multiple species grazing [41–43]. Cattle are, by far, the grazing animals of choice in South Dakota (1.8 million cattle vs. 260,000 sheep) [44] and across the Northern Great Plains of the US.

Herbicide applications are reported to be the most effective methods for *Artemisia absinthium* control [22, 45–47]. There are numerous reports about the enhanced effectiveness of combining weed control strategies for weed suppression in grazing lands [1, 7, 47–49]. In this study, using 2,4-D ester herbicide in combination with grazing, helped remove *Artemisia absinthium* growth for the first growing season. Some herbicides affect the palatability of certain plants, encouraging livestock to eat plants they would normally avoid, like poisonous plants [50]. However, precautions must be taken if spraying 2,4-D [24] because this herbicide can cause plants to accumulate excess nitrate, become more palatable, and result in nitrate poisoning of livestock [51]. There are a few grazing restrictions for 2,4-D ester [24]. For example, meat animals could be grazed immediately after application, but not within 7 days of slaughter; and restrictions for a dairy animals differed with no grazing within 7 days post-application.

5. Conclusions

Healthy rangelands grow more grass which aids in *Artemisia absinthium* control by preventing infestations and providing competition to newly establishing plants. Grass density can be optimized by managing livestock to minimize overgrazing through rotational grazing or avoiding heavy, early season grazing [22]. Based on *Artemisia absinthium* size increase in the 2014 recovery area after the early spring rotational grazing/summer rest, it appears that rotational grazing later in the growing season (as in 2013) achieved better suppression of *Artemisia absinthium* patches, although cattle did not necessarily consume *Artemisia absinthium*.

Once present, our study showed that grazing provided temporary reductions to *Artemisia absinthium* patches, with greater reductions in the mob-grazed and rotational/spray treatments than the rotational grazed treatment. Shoots of smaller plants and those in smaller patches appeared to be consumed in both mob grazing and rotational grazing when 2,4-D ester was applied. However, even the most decimated plants had shoots the following season. Once pastures are infested, long-term management plans are needed to keep *Artemisia absinthium* in check.

We found that mob grazing with cattle for 12 or 24 h in pastures where *Artemisia absinthium* was present did indeed improve *Artemisia absinthium* control of smaller plants (as measured in plant volume) with concomitant high forage utilization. Rotational grazing at lower stocking rates for 20 days (late-May through mid-June), when combined with 2,4-D application, also suppressed *Artemisia absinthium* for that growing season. Early (mid-April) rotational grazing with a summer rest resulted in much larger *Artemisia absinthium* plants and patches in the fall. We could not verify the statements that mob grazing would result in (1) an increase of two or more cm of soil per year, nor (2) a species composition change due to the intense grazing, which are two positive benefits of mob grazing often discussed in trade journal articles [15, 19, 52]. In addition, we did not assess the impact of mob grazing on animal performance, although in a single one-time grazing situation, a change in this parameter would not be expected. Long term management plans are needed for *Artemisia absinthium*, as all *Artemisia absinthium* patches observed after the first grazing season produced shoots the year following grazing, regardless of the amount of grazing or trampling damage that was sustained.

Acknowledgements

Thanks to Mr. R. Smith, Hayti, SD for providing access to land and cattle. Funding support for this study was provided by the South Dakota Agricultural Experiment Station and USDA, NRCS Conservation Innovation Grant (CIG) 3FH560 "Demonstrating Mob Grazing Impacts in the Northern Great Plains on Grazing Land Efficiency, Botanical Composition, Soil Quality, and Ranch Economics."

Nomenclature

Cattle *Bos taurus* L.

Absinth wormwood *Artemisia absinthium* L.

Author details

Heidi Reed[1*], Alexander Smart[2], David E. Clay[1], Michelle Ohrtman[1] and Sharon A. Clay[1]

*Address all correspondence to: hxm5183@psu.edu

1 South Dakota State University Department of Agronomy, Horticulture and Plant Science, Brookings, South Dakota, USA

2 South Dakota State University Department of Natural Resource Management, Brookings, South Dakota, USA

References

[1] DiTomaso JM. Invasive weeds in rangelands: Species, impacts, and management. Weed Science. 2000;**48**:255-265

[2] Pimentel D, Zuniga R, Morrison D. Update on the environmental and economic costs associated with alien-invasive species in the United States. Ecological Economics. 2005;**52**:273-288

[3] Popay I, Field R. Grazing animals as weed control agents. Weed Technology. 1996;**10**: 217-231

[4] Plant Protection Act. 7 U.S.C. 7701 et seq. 114 stat. 438 [Internet]. 2000. Available from: http://www.aphis.usda.gov/plant_health/plant_pest_info/weeds/downloads/PPAText.pdf [Accessed: April 13, 2018]

[5] Ohlenbusch PD, Towne G. Rangeland Weed Management. KState Bulletin, MF-1020 [Internet]. 1991. Available from: www.bookstore.ksre.ksu.edu/pubs/mf1020.pdf [Accessed: April 13, 2018]

[6] Sellers BA, Ferrell JA. Weed Management in Pastures and Rangeland—2017. University of Florida IFAS Extension [Internet]. 2017. Available from: SS-AGR-08.edis.ifas.ufl.edu/wg006 [Accessed: June 30, 2017]

[7] Frost RA, Launchbaugh KL. Prescription grazing for rangeland weed management: A new look at an old tool. Rangelands. 2003;**25**:43-47

[8] Holechek JL. Comparative contribution of grasses, forbs, and shrubs to the nutrition of range ungulates. Rangelands. 1984;**6**:261-263

[9] Bailey DW, Brown JR. Rotational grazing systems and livestock grazing behavior in shrub-dominated semi-arid and arid rangelands. Rangeland Ecology Management. 2011;**64**:1-9

[10] Senft RL, Rittenhouse LR, Woodmansee DRG. Factors influencing patterns of cattle grazing behavior on shortgrass steppe. Journal of Range Management. 1985;**38**:82-87

[11] Fishel F. Plants Poisonous to Livestock. Columbia, MO: University of Missouri Extension; 2001. p. G-4970

[12] Crider FJ. Root-growth stoppage resulting from defoliation of grass. U.S. Dept. Agr. Tech. Bull. 1102.1955.23 pg

[13] Frost WE, Smith EL, Ogden PR. Utilization guidelines. Rangelands. 1994;**6**:256-259

[14] Savory A, Butterfield J. A commonsense revolution to restore our environment. In: Holistic Management. 3rd ed. Washington, DC: Island Press; 2016. 552pp. ISBN: 9781610917438

[15] Gordon K. Mob Grazing 101. Herford World [Internet]. 2011. Available at: www.Herford.org; hereford.org/static/files/0111_MobGrazing.pdf [Accessed: December 15, 2017]

[16] Barnes MK, Norton BE, Maeno M, Malechek JC. Paddock size and stocking density affect spatial heterogeneity of grazing. Rangeland Ecology & Management. 2008;**61**:380-388

[17] Pleasants AB, Shorten PR, Wake GC. The distribution of urine deposited on a pasture from grazing animals. Journal Agricutral. Sciences. 2006;**145**:81-86

[18] Frank DA, McNaughton SJ. Evidence for the promotion of aboveground grassland production by native large herbivores in Yellowstone National Park. Oecologia. 1993;**96**:157-161

[19] Thomas HS. Ranchers Sing the Praises of Mob Grazing of Cattle. Beef Magazine [Internet] 2012. February 28, 2012. Available from: http://www.beefmagazine.com/pasture-range/ranchers-sing-praises-mob-grazing-cattle [Accessed: March 03, 2018]

[20] McCartney DH, Bittman S. Persistence of cool-season grasses under grazing using the mob-grazing technique. Canadian Journal of Plant Science. 1994;**74**:723-728

[21] Senft RL. Hierarchical foraging models: Effects of stocking and landscape composition on simulated resource use by cattle. Ecological Modeling. 1989;**46**:283-303

[22] Maw MG, Thomas AG, Stahevitch A. The biology of Canadian weeds. 66. *Artemisia absinthium* L. Canadian Journal of Plant Science. 1985;**65**:389-400

[23] Myer H. Mob grazing as a perennial weed management tool in South Dakota grazing-lands [MS thesis]. Brookings, SD: SDSU; 2015, p. 141

[24] Anonymous. 2, 4-D Ester 4. Available from: http://www.cdms.net/LDat/ld 40L006.pdf [Accessed: November 14, 2014]

[25] Harmoney KR, Moore KJ, George JR, Brummer EC, Russell JR. Determination of pasture biomass using four indirect methods. Agronomy Journal. 1997;**89**:665-672

[26] Barnhart SK. Estimating Available Pasture Forage [Internet]. Iowa State Univ. Extension. 2009. PM 1758. https://store.extension.iastate.edu/Product/Estimating-Available-Pasture-Forage-PDF

[27] Grazing Stick Instruction Manual. The Samuel Roberts Nobel Foundation [Internet]. 2007. Available at: http://www.noble.org/ag/forage/grazingstick/index.html [Accessed: April, 2018]

[28] Steel RGD, Torrie JH. Principles and Procedures of. Statistics. 2nd ed. New York: McGraw-Hill, Inc; 1980. 633p. ISBN: 0-07-060926-8

[29] Anderson VJ, Briske DD. Herbivore-induced species replacement in grasslands: Is it driven by herbivory tolerance or avoidance? Ecological Applications. 1995;**5**:1014-1024

[30] Fuhlendorf SD, Engle DM. Restoring heterogeneity on rangelands: Ecosystem management based on evolutionary grazing patterns. Bio Science. 2001;**51**:625-632

[31] Manske LL. General Description Of Grass Growth And Development And Defoliation Resistance Mechanisms. Range Management Report DREC 98-1022. Dickinson, North Dakota: NDSU Dickinson Research Extension Center; 1998. 12p

[32] Nishica T, Shimizu N, Ishida M, Onoue T, Harashima N. Effect of cattle digestion and of composting heat on weed seeds. Japan Agricultural. Research Quarterly. 1998;**32**:55-60

[33] Rahimi S, Mashhadi HR, Banadaky MD, Mesgaran MB. Variation in Weed Seed Fate Fed to Different Holstein Cattle Groups. PLoS ONE. 2016;**11**(4):e0154057. https://doi.org/10.1371/journal.pone.0154057

[34] Stevens OA. Flowering dates of weeds in North Dakota. North Dakota Agricultural Experiment Station Bimonthly. Bulletin. 1956;**18**:209-213

[35] Kohl MT, Krausman PR, Kunkel K, Williams DM. Bison versus cattle: Are they ecologically synonymous? Rangeland Ecology Management. 2013;**66**:721-731

[36] Harker KN, Baron VS, Chanasyk DS, Naeth MA, Stevenson FC. Grazing intensity effects on weed populations in annual and perennial pasture systems. Weed Science. 2000;**48**:231-238

[37] Hart RH, Clapp S, Test PS. Grazing strategies, stocking rates, and frequency and intensity of grazing on western wheatgrass and blue grama. Journal of Range Management. 1993;**46**:122-126

[38] Isbell F, Calcagno V, Hector A, Connolly J, Harpole WS, Reich PB, Scherer-Lorenzen M, Schmid B, Tilman D, van Ruijven J, Weigelt A, Wilsey BJ, Zavaleta ES, Loreau M. High plant diversity is needed to maintain ecosystem services. Nature. 2011;**477**:199-202

[39] Deak A, Hall MH, Sanderson MA. Grazing schedule effect on forage production and nutritive value of diverse forage mixtures. Agronomy Journal. 2009;**101**:408-414

[40] Laliberte E, Lambers H, Norton DA, Tylianakis JM, Huston MA. A long-term experimental test of the dynamic equilibrium model of species diversity. Oecologia. 2013;**171**:439-448

[41] Dabaan ME, Magadlela AM, Bryan WB, Arbogast BL, Prigge EC, Flores G, Skousen JG. Pasture development during brush clearing with sheep and goats. Journal of Range Management. 1997;**50**:217-221

[42] Sahlu T, Dawson LJ, Gipson TA, Hart SP, Merkel RC, Puchala R, Wang Z, Zeng S, Goetsch AL. ASAS Centennial Paper: Impact of animal science research on United States goat production and predictions for the future. Journal of Animal Science. 2009;**87**:400-418

[43] Lupton CJ. ASAS Centennial Paper: Impact of animal science research on United States sheep production and predictions for the future. Journal of Animal Science. 2008;**86**:3252-3274

[44] 2017 State Agriculture Overview, South Dakota [Internet]. Available from: https://www.nass.usda.gov/Quick_Stats/Ag_Overview/stateOverview [Accessed: April 12, 2018]

[45] Carey JH. *Artemisia absinthium*. In: Fire Effects Information System [Internet]. USDA, Forest Service, Rocky Mountain Research Station, Fire Sciences Laboratory. 1994. Available from: www.invasive.org/weedcd/pdfs/feis/Artemisiaabsinthium.pdf [Accessed: November 22, 2017]

[46] Eckardt N. Element Stewardship Abstract for *Artemisia absinthium*. Arlington, Virginia: The Nature Conservancy. 1987. Available from: www.invasive.org/weedcd/pdfs/tnc-weeds/arteabs.pdf [Accessed: November 14, 2017]

[47] Lym R, Messersmith C, Dexter A. Absinth Wormwood Control [Internet]. North Dakota State University Extension. 2013. W-838

[48] Lacey JR, Sheley RL. Leafy spurge and grass response to picloram and intensive grazing. Journal of Range Management. 1996;**49**:311-314

[49] Lym RG, Sedivec KK, Kirby DR. Leafy spurge control with angora goats and herbicides. Journal of Range Management. 1997;**50**:123-128

[50] Moechnig M, Deneke DL, Wrage LJ, Rosenberg M. Weed Control: Pasture and Range [Internet]. South Dakota State University Extension Service. 2013

[51] Holechek JL, Pieper RD, Herbel CH. Range Management: Principles and Practices. 6th ed. Upper Saddle River, NJ: Pearson/Prentice Hall; 2004. 607p. ISBN-13: 978-0130474759

[52] Hancock D. [Internet] Is Mob Grazing as Effective as we Thought? On Pasture. Availableat: https://onpasture.com/2017/11/06/is-mob-grazing-as-effective-as-we-thought/2017 [Accessed: April 12, 2018]

Chapter 5

Domestic Livestock and Its Alleged Role in Climate Change

Albrecht Glatzle

Additional information is available at the end of the chapter

http://dx.doi.org/10.5772/intechopen.80389

Abstract

It is very old wisdom that climate dictates farm management strategies. In recent years, however, we are increasingly confronted with claims that agriculture, livestock husbandry, and even food consumption habits are forcing the climate to change. We subjected this worrisome concern expressed by public institutions, the media, policy makers, and even scientists to a rigorous review, cross-checking critical coherence and (in)compatibilities within and between published scientific papers. Our key conclusion is there is no need for anthropogenic emissions of greenhouse gases (GHGs), and even less so for livestock-born emissions, to explain climate change. Climate has always been changing, and even the present warming is most likely driven by natural factors. The warming potential of anthropogenic GHG emissions has been exaggerated, and the beneficial impacts of manmade CO_2 emissions for nature, agriculture, and global food security have been systematically suppressed, ignored, or at least downplayed by the IPCC (Intergovernmental Panel on Climate Change) and other UN (United Nations) agencies. Furthermore, we expose important methodological deficiencies in IPCC and FAO (Food Agriculture Organization) instructions and applications for the quantification of the manmade part of non-CO_2-GHG emissions from agro-ecosystems. However, so far, these fatal errors inexorably propagated through scientific literature. Finally, we could not find a clear domestic livestock fingerprint, neither in the geographical methane distribution nor in the historical evolution of mean atmospheric methane concentration. In conclusion, everybody is free to choose a vegetarian or vegan lifestyle, but there is no scientific basis, whatsoever, for claiming this decision could contribute to save the planet's climate.

Keywords: greenhouse gas emissions, carbon dioxide, methane, nitrous oxide, agro-ecosystems, deforestation, climate change

1. Introduction

Since its early origins, mankind adapts to the prevailing climatic conditions (from the arctic to the tropical rainforest) and copes fairly successfully with natural climate variability. It is very old wisdom that climate dictates farm management strategies. Fairly new, however, is the idea that agriculture, livestock husbandry, and food consumption habits are forcing supposedly the climate to change. This idea spread across the globe when thousands of media reports picked up the central message of the famous FAO report "Livestock's Long Shadow" [1], which blamed domestic livestock of causing serious environmental hazards such as climate change, through greenhouse gas (GHG) emissions. Another FAO report [2] basically transmitted the same message, reducing, however, somewhat the livestock contribution to global GHG emissions from 18 to 14.5%. But dramatic figures of emission intensity were still maintained particularly for South American pasture-based beef production (**Figure 1**).

The worrisome messages launched by the FAO were eagerly disseminated by several environmentalist and even ecclesiastic organizations. They also triggered political action: there was a public audience in the European Parliament in November 2009 about the topic "Less Meat = Less Heat." And at the Conference of Partners in Paris COP21 in 2015, this topic was also subject in the climate negotiations. And even in scientific literature, reduction of livestock numbers and meat consumption was recommended [3]. These concerns expressed by public institutions, the media, politics, and even science evoke the question: is global climate really at risk from livestock husbandry and cropping?

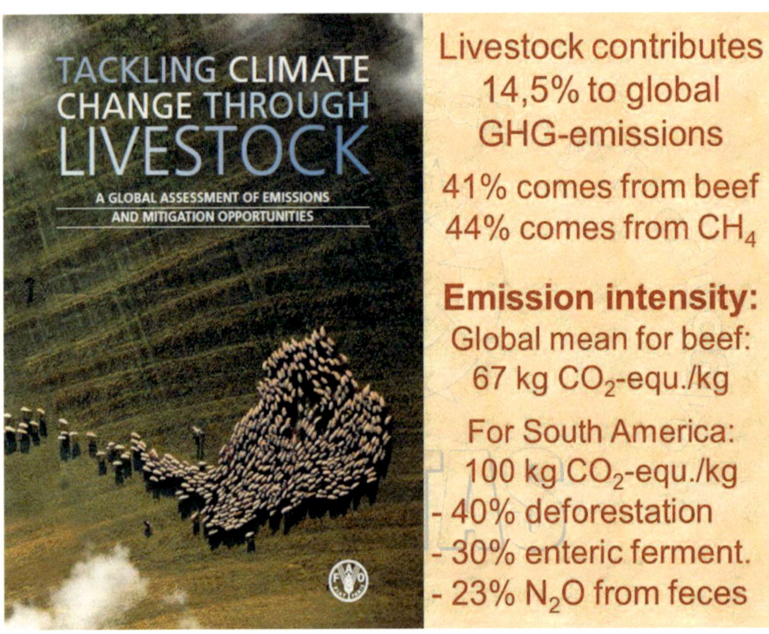

Figure 1. Key conclusions from Gerber et al. [2].

2. Methodological procedure

To answer this question, we did extensive review work, cross-checking critically coherence and (in)compatibilities between several published papers and data, and came to distinct results to what one would expect when listening to environmentalists and political climate change activists.

3. Results and discussion

3.1. About GHG emissions in the context of livestock husbandry

3.1.1. Carbon dioxide (CO_2)

CO_2 emitted by human consumption of cereals, meat, and milk, by livestock respiration and forage digestion, does not increase atmospheric CO_2 levels, as this is part of the natural carbon cycle. Not a single human- or livestock-born CO_2 molecule is additionally released into the atmosphere, as it has previously been captured through photosynthesis. The amount of CO_2 released annually by humans and livestock is offset by regrowing CO_2-assimilating forages and crops. The only sources of *additional* CO_2 emissions caused by agriculture and livestock husbandry, beyond the natural carbon cycle, are:

- fossil fuel consumption during production, processing, and marketing, such as transportation, soil tillage, harvesting, and fertilizer manufacturing,
- deforestation for reclamation of pasture and cropland, and
- soil organic matter decomposition from degrading grasslands and arable lands, as determined by the difference of ecosystemic carbon stocks before and after certain human interventions.

Usage of fossil fuels is considerable in industrial livestock production systems which rely on forage cropping and feed transportation to the confined animals. In grazing systems, however, fuel consumption is rather low. Fossil fuel-related emission intensity of feed is less than 0.05 CO_2 kg^{-1} of dry matter intake in grazing systems and around 0.3 in feedlots [4]. The widespread perception that only feedlot intensification can reduce the overall GHG emission intensity (per kg of beef produced) was recently challenged by Paige et al. [5] who found considerable soil organic carbon sequestration in certain grazing systems which even offset methane emissions from enteric fermentation. However, after any sort of land use change, the rate of soil carbon sequestration or of carbon loss is changing over time until a new equilibrium level is reached for each kind of land management [6].

Deforestation for pasture establishment causes a unique one-time CO_2 release from burning and decomposition of woody vegetation. For emission intensity calculations, deforestation-born emissions have to be shared out over the accumulated animal products generated during the total utilization period of the very pasture, which replaced the forest. This may easily be hundreds of years (as in the case of European grasslands). In the long run, total production accumulates to huge quantities and the deforestation part of the emission intensity (CO_2 emitted per kg of carcass weight) approaches zero (**Figure 2**).

Unfortunately, in published literature, emissions from deforestation are treated inconsistently. They are either neglected or charged entirely to the year of their appearance onto a product which is not necessarily related to the ongoing deforestation, such as total beef production in South America (e.g., **Figure 1**). For Europe, however, these emissions are usually ignored as they took place 500 years and longer ago.

In spite of ongoing deforestation, world vegetation cover, particularly in (semi-)arid regions, has improved in the past 30 years due to rising CO_2, as a satellite image-based analysis by CSIRO Australia [7] and Geoscience Institutes in Denmark and Spain [8] has shown. Another study of 32 authors from 24 institutions from 8 countries, published on the NASA website, found a significant increase in the leaf area index on most of the earth's vegetated surface, during the past 35 years, for which increasing CO_2 emissions are considered responsible at a 70% level [9, 10].

In the Northern Hemisphere with big landmasses covered with vegetation, the annual oscillation of CO_2 rose considerably in the past decades. In 2013, 36% more CO_2 was captured in spring and summer and released again in wintertime than 45 years ago. The growing annual amplitude with more CO_2 in the air is a clear indicator of a tremendous vegetation response to increased CO_2 levels [11]. Fully in line with this finding is another paper published in *Nature* providing evidence that twentieth-century CO_2 emissions caused an over 30% increase in Global Terrestrial Gross Primary Production [12].

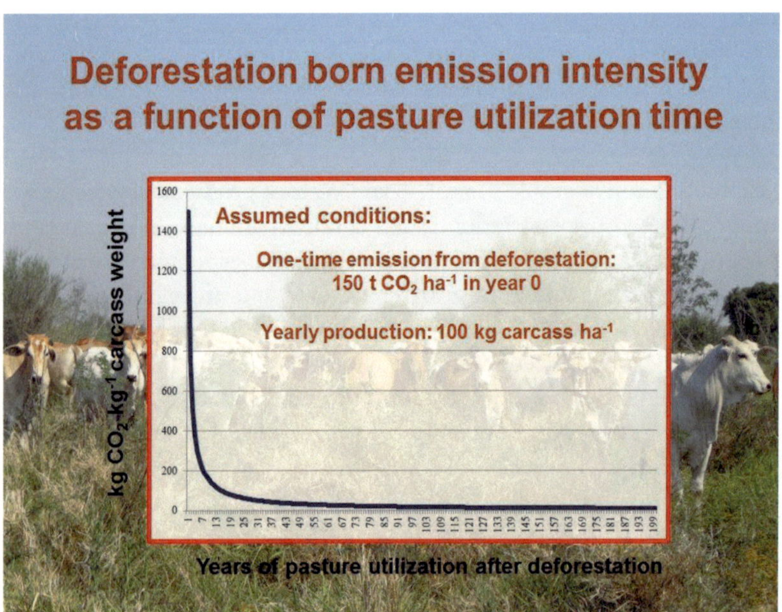

Figure 2. Modeling deforestation-born emission intensity (kg CO_2 emitted per kg of carcass weight produced).

Former IPCC author and reviewer Indur Goklany [13] estimated the global fertilization value of manmade CO_2 in the atmosphere to 140 billion US$ every year. Therefore, anthropogenic CO_2 contributes considerably to global food security. There are dozens of studies corroborating the efficiency of CO_2 as a fertilizer of our crops, pastures, and forests [14]. Nevertheless, UNEP projects (United Nations Environmental Program) such as the initiative TEEB (The Economy of Ecosystems and Biodiversity for Agriculture and Food) categorically ignore the obvious beneficial effects of manmade CO_2 emissions in their economic assessments. So do the authors of a recent assessment of potential economic damages under UN mitigation targets [15]. The well-established desirable effects of manmade CO_2 are entirely disregarded, whereas the global warming thresholds of future emission scenarios, as proposed by the IPCC, are fully accepted and related to potential economic losses, differentiated by regions. However, this widely accepted approach does not represent an objective and trustworthy method (see Chapter 3.2).

During most of the geological eras, atmospheric CO_2 concentrations were higher than today. At the last glaciation maximum, however, 18,000 years ago, CO_2 concentration reached as little as 180 ppm, low enough to stunt plant growth [16]. Therefore, quite a number of authors celebrate the recirculation of CO_2 by fossil fuel burning to secure long-time survival of life on earth. Taking into account that CO_2 is essential nutrient for life, is the only carbon source of all biomass, is fertilizing our crops and pastures, and is greening our deserts as it improves water use efficiency and therefore drought resistance of plants [17], this trace compound in the air (0.04% vol.) qualifies for being the most important, however limiting, nutrient for life. It is not the air pollutant as which it is seemingly exposed in the media and even by members of the scientific community. CO_2 is a transparent and odorless trace gas of which we are respiring about 5 kg every day.

3.1.2. Non-CO_2 GHGs: methane (CH_4) and nitrous oxide (N_2O)

Other agricultural GHGs such as methane (CH_4) and nitrous oxide (N_2O) also form part of natural cycles, just like CO_2. An easily understandable overview on methane and nitrous oxide dynamics in the atmosphere has been worked out by Stephen Zwick in LA Chefs Column [18]. There are natural and manmade sinks and sources for CH_4 and N_2O (**Figure 3**); there is, however, some confusion in the quantification of the manmade part of their emissions from agro-ecosystems. The IPCC Guidelines for National Greenhouse Gas Inventories [19] meticulously provide instructions, emission factors, and formulas to estimate the emissions from the various sources in managed ecosystems. Emissions from pristine or native ecosystems are explicitly not taken into account, as they are not manmade. However, all managed agro-ecosystems replaced native ecosystems at some stage in history which also had been sources of considerable methane and nitrous oxide emissions.

In order to get the effective manmade part of the emissions from managed ecosystems, one has to subtract the baseline emissions of the respective native ecosystems or of the pre-climate change-managed ecosystems from those of today's agro-ecosystems (**Figure 4**). Omitting this correction leads to a systematic overestimation of farm-born non-CO_2 GHG emissions. Scientific publications generally do not take this consideration into account, as farm-born CH_4 and N_2O emissions are consistently interpreted at a 100% level as an *additional* anthropogenic GHG source, just like fossil fuel-born CO_2. As the mentioned IPCC guidelines [19] are taken for the ultimate reference, this severe methodological deficiency propagated through scientific literature.

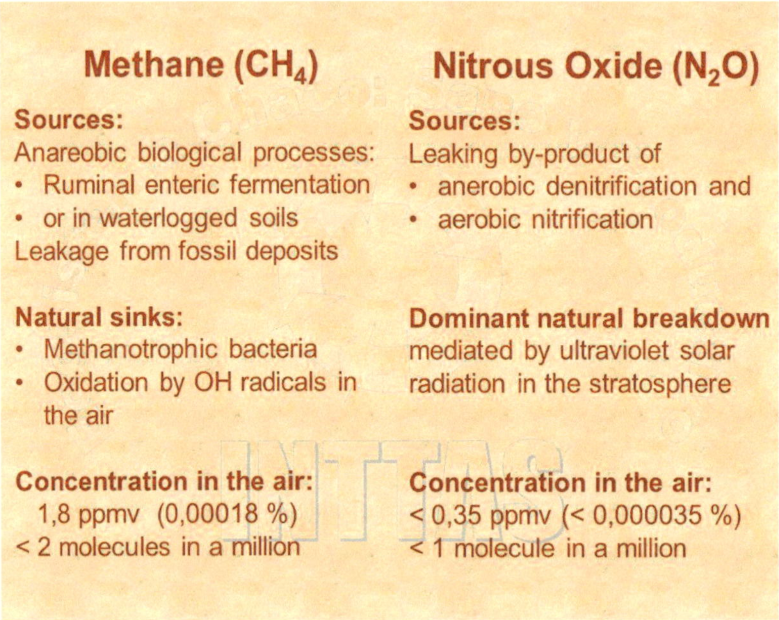

Figure 3. Natural and anthropogenic sources and sinks of the non-CO_2 GHGs methane and nitrous oxide.

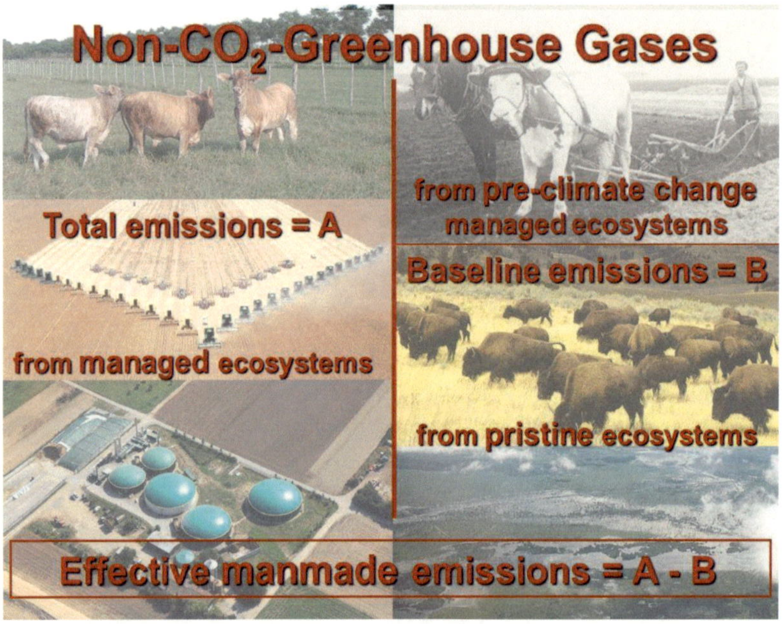

Figure 4. How to estimate correctly manmade non-CO_2 GHG emissions from agro-ecosystems.

Temporarily waterlogged or flooded pristine ecosystems or those with a high density of wild ungulates might have emitted the same amount or even more methane per hectare and year than they did after land reclamation and utilization. So net anthropogenic methane emissions from certain agro-ecosystems could be zero or even assume a negative value.

The same applies to nitrous oxide, particularly in farming systems where no or little synthetic nitrogen fertilizer is used such as most pastoral systems: ecosystem management and herbage consumption by livestock might increase somewhat the turnover rate of nitrogen but does not increase the quantity of nitrogen in circulation from which N_2O is emitted as a by-product from nitrification and denitrification.

Dung patches concentrate the nitrogen ingested from places scattered across the pasture. Nichols et al. [20] found no significant differences between emission factors from the patches and the rest of the pasture, which means the same amount of nitrous oxide is emitted whether or not the herbage passes livestock's intestines. However, the IPCC and FAO do consider mistakenly all nitrous oxide leaking from manure as livestock-born and therefore manmade.

Comparing, for instance, sown grassland with native bushland in the Gran Chaco, which contains many leguminous species, it becomes evident that nitrogen stocks are higher and more nitrogen is circulated annually in native bushland than in sown pasture (**Figure 5**). Therefore, in spite of the presence of grazing animals in the grassland, there is likely more nitrous oxide produced from bushland than from grassland after bush clearing and pasture establishment.

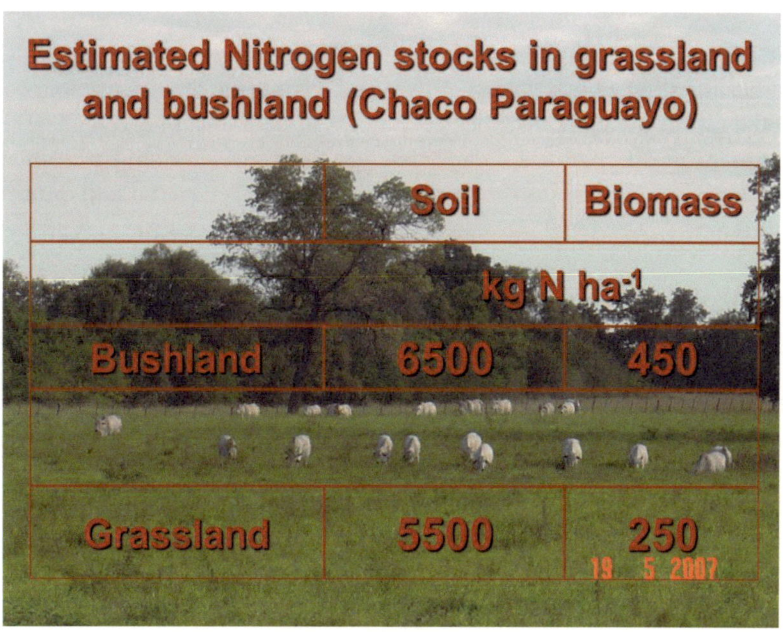

Figure 5. Ecosystemic nitrogen stocks in grassland and bushland (Chaco, Paraguay).

Hence, instead of charging the emission intensity of South American Beef with 23 kg of CO_2-equ. kg^{-1} of CW (carcass weight) for nitrous oxide emissions from animal feces (**Figure 1**), there should rather be a negative value when corrected for the emissions from the respective pre-land use pristine ecosystem. Similar thoughts can be made for the enteric fermentation and deforestation part of emission intensity charges.

3.1.3. Global methane emissions and livestock

The rise of methane emissions beginning around 1850 coincides perfectly with the progressive use of fossil energy. But the methane growth rate fell to zero at the turn of the millennium as shown by Quirk [21], cited from [22]. The stabilization of methane emissions in the 1990s is very likely associated with the adoption of modern technology in fossil fuel production and use, particularly the replacement of leaking pipelines in the former Soviet Union [21].

Between 1990 and 2005, the world cattle population rose by more than 100 million head (according to FAO statistics). During this time, atmospheric methane concentration stabilized completely. These empirical observations show that livestock is not a significant player in the global methane budget [23]. This appreciation has been corroborated by Schwietzke et al. [24] who suggested that methane emissions from fossil fuel industry and natural geological seepage have been 60–110% greater than previously thought.

When looking to the global distribution of average methane concentrations as measured by ENVISAT (Environmental Satellite) [25] and the geographical distribution of domestic animal density, respectively [1], no discernible relationship between both criteria was found [22].

Although the most recent estimates of yearly livestock-born global methane emissions came out 11% higher than earlier estimates [26], we still cannot see any discernible livestock fingerprint in the global methane distribution (**Figure 6**). The idea of a considerable livestock contribution to the global methane budget relies on theoretical bottom-up calculations. Even in recent studies, e.g., [27], just the emissions per animal are measured and multiplied by the number of animals. Ecosystemic interactions and baselines over time and space are generally ignored [28]. Although quite a number of publications, such as the excellent most recent FCRN report (Food Climate Research Network) [29], do discuss extensively ecosystemic sequestration potentials and natural sources of GHGs, they do not account for baseline emissions from the respective native ecosystems when assessing manmade emissions of non-CO_2 GHGs from managed ecosystems. This implies a systematic overestimation of the warming potential, particularly when assuming considerable climate sensitivity to GHG emissions. However, even LA Chefs Column [18], in spite of assuming a major global warming impact of methane, came to the conclusion: "When methane is put into a broader rather than a reductive context, we all have to stop blaming cattle ('cows') for climate change."

3.2. About the climate response to manmade GHG emissions

Having shown considerable beneficial effects of manmade CO_2 emissions on nature, agriculture, and global food security and having shown severe IPCC and FAO deficiencies in the quantification of the manmade part of non-CO_2 GHG emissions, we need to have a closer look to the alleged evil human emissions of natural GHGs are accused of: causing climate change through global warming.

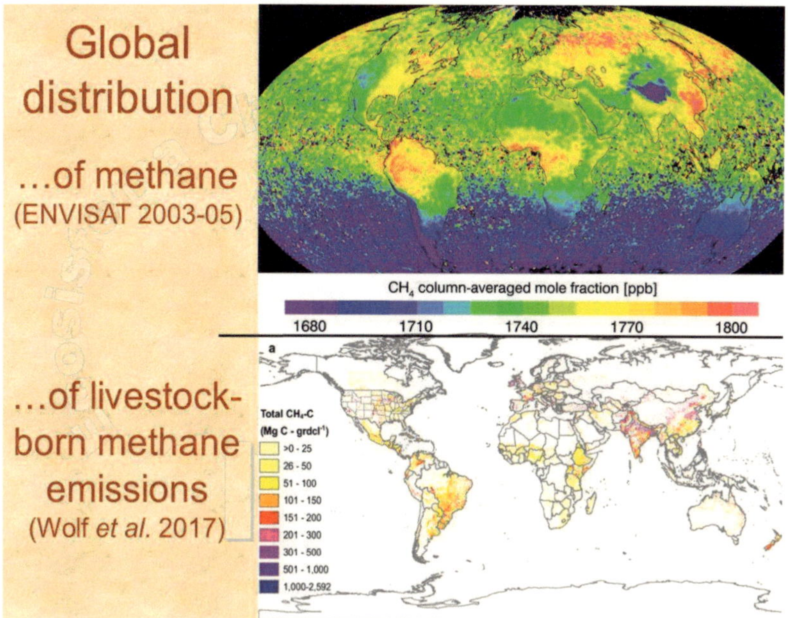

Figure 6. Domestic livestock-born methane emissions are of negligible importance for the global geographical methane distribution [25, 26].

There is, however, a growing divergence between observed and modeled temperatures. In spite of steadily increasing CO_2 levels, observed temperatures are ways below most published temperature projections (**Figure 7**).

Critical scientists are not surprised of this reality, showing that model validation has pitiably failed. In Table 2.11 of the Fourth IPCC Assessment Report AR4 [30], 16 variables were identified as global warming-forcing agents and used for modeling. The level of understanding for 11 of them is specified as "low to very low." Under such premises, reliable modeling is impossible. Yet the IPCC comes up with a 90–95% certainty that human activity has been the main single driver of the slight warming observed during the past century.

According to Gervais [11], published estimates of climate sensitivity to CO_2, as defined as temperature rise with CO_2 doubling, are in rapid decline since the turn of the millennium. The logical implication of this finding is that, in the past, climate models systematically exaggerated temperature projections into the future. Moreover, for the time between 1993 and 2015, when about 40% of total CO_2 was emitted since the beginning of the industrial revolution, Gervais could not find any discernible correlation between atmospheric concentration of CO_2 and mean global temperature anomaly in the low stratosphere (as measured by satellites), where according to the radiative-convective models, the most marked signature of temperature change was predicted [11]. Recent investigations support the idea of biases in IPCC climate model simulations, most of which show spurious warming associated with its alleged impacts such as glacier melting and sea level rise [32–36].

Furthermore, a growing number of peer-reviewed papers give evidence of pronounced warm periods during the Holocene, since the end of the last ice age, 10,000 years ago, in spite of the preindustrial atmospheric CO_2 levels in those times [28]. Gernot Patzelt from Innsbruck

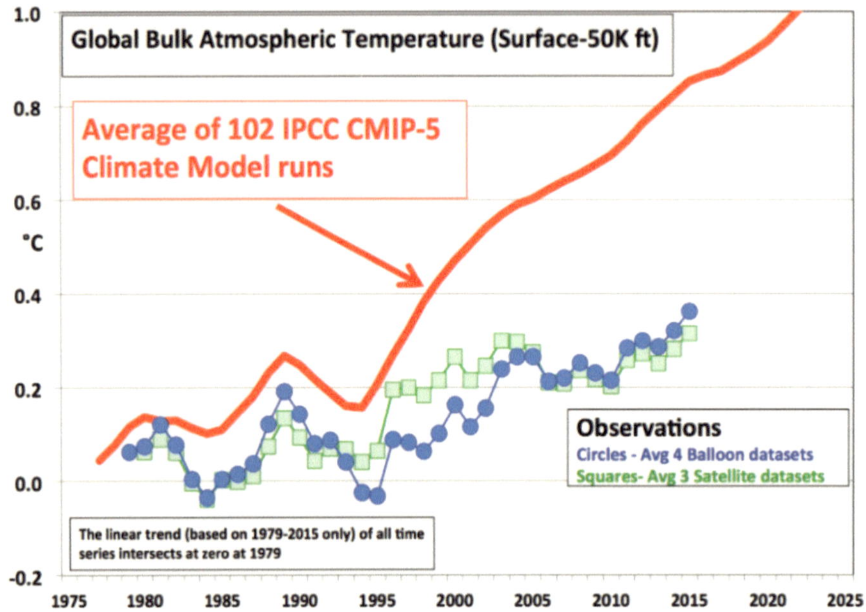

Figure 7. Midtropospheric temperature variations: observations (by satellite and balloons) versus IPCC models [31].

University [37] recovered ancient tree trunks conserved in moors and glaciers well above the present day tree lines, all across the Alps (**Figure 8**).

Patzelt irrefutably concluded that 65% of the Holocene summer temperatures had been warmer than today because the tree lines were at higher altitudes than today. Other studies

Figure 8. These tree trunks uncovered from retreating glaciers are irrefutable witnesses of extended preindustrial warm periods as they grew up well above the present-day tree lines [38].

from stalagmites in the Alps [39] and tree line investigations in Lapland [40] gave similar results, just as did ice core analyses from Greenland [41] and from the Antarctica [42].

The IPCC faces considerable problems of explaining the numerous preindustrial warm periods: among the radiative forcing components as published in the latest IPCC report in 2013 [43], anthropogenic CO_2, methane, and nitrous oxide emissions are represented with prominent bars and hence are supposed to be the key drivers of global warming. On the other hand, the solar influence has been reduced to a tiny effect, just representing the observed small variation of direct solar irradiation (**Figure 9**).

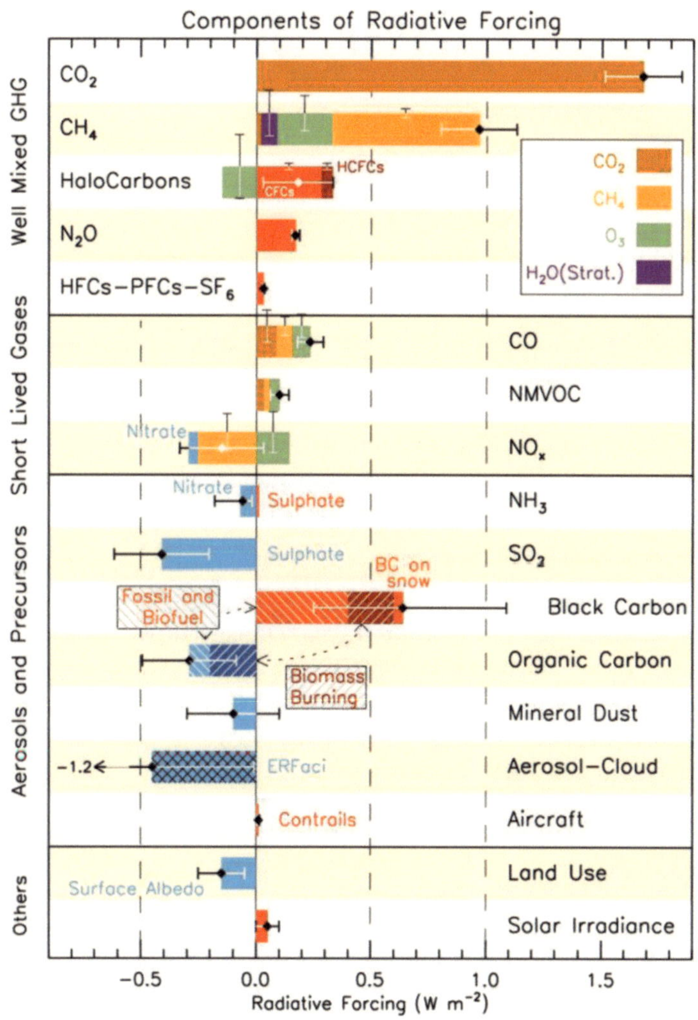

Figure 9. Natural and anthropogenic global warming forcing agents as defined and quantified by the IPCC (Figures 8-17 from [43]). These are incompatible with the well-documented prominent warm periods, which occurred in spite of preindustrial CO_2 levels.

These global warming forcing agents defined by the IPCC [43] obviously ignore the potent indirect solar influences produced by solar magnetic activity associated with sunspot occurrence. Lockwood et al. [44] clearly showed the relevance of solar activity indicators for the heliospheric cosmic ray modulation potential and the associated cooling and warming of the earth during the past 400 years. The causal chain between solar magnetic activity, cosmic ray flux hitting the earth, cloud formation potential, and mean global temperature has been shown by Svensmark and Friis-Christensen [45] and was convincingly defended against premature critics [46].

4. Conclusion

There is no need for anthropogenic emissions of GHGs, and even less so for livestock-born emissions, to explain climate change. When looking closely to published scientific data and facts, we conclude that

- eternal climate change, also the present one, is most likely driven by natural factors,
- the warming potential of anthropogenic GHGs has very likely been exaggerated by the IPCC and the media, and
- beneficial impacts of anthropogenic CO_2 emissions for nature, agriculture, and global food security have been systematically ignored.

Furthermore, we exposed important methodological deficiencies in IPCC and FAO instructions and applications for the quantification of the manmade part of non-CO_2 GHG emissions from agro-ecosystems. Finally, we could not find a domestic livestock fingerprint, neither in the geographical methane distribution nor in the historical evolution of the atmospheric methane concentration.

Consequently, in science, politics, and the media, climate impact of anthropogenic GHG emissions has been systematically overstated. Livestock-born GHG emissions have mostly been interpreted isolated from their ecosystemic context, ignoring their negligible significance within the global balance. There is no scientific evidence, whatsoever, that domestic livestock could represent a risk for the Earth's climate.

Acknowledgements

The author wants to thank Prof. Emeritus Dr. Nils-Axel Mörner, Stockholm University, for inspiring ideas, stimulating conversations, and providing with relevant references.

Conflict of interest

The author has no conflict of interest to declare.

Author details

Albrecht Glatzle

Address all correspondence to: albrecht.glatzle@gmail.com

INTTAS (Initiative for Research and Extension of Sustainable Agrarian Technologies), Filadelfia, Paraguay

References

[1] Steinfeld H, Gerber P, Wassenaar T, Castel V, Rosales M, de Haan C. Livestock's Long Shadow. Rome: LEAD (The Livestock, Environment and Development Initiative), FAO; 2006. http://www.fao.org/docrep/010/a0701e/a0701e00.HTM

[2] Gerber PJ, Steinfeld H, Henderson B, Mottet A, Opio C, Dijkman J, et al. Tackling Climate Change through Livestock—A Global Assessment of Emissions and Mitigation Opportunities. Rome: FAO; 2013:15-17, 24, 68-69. http://www.fao.org/3/a-i3437e.pdf

[3] Ripple WJ, Smith P, Haberl H, Montzka SA, McAlpine C, Boucher DH. Ruminants, climate change and climate policy. Nature Climate Change. 2014;**4**:2-5. DOI: 10.1038/nclimate2081. https://www.nature.com/articles/nclimate2081

[4] Opio C, Gerber P, Mottet A, Falcucci A, Tempio G, MacLeod M, Vellinga T, Henderson B, Steinfeld H. Greenhouse Gas Emissions from Ruminant Supply Chains—A Global Life Cycle Assessment. Rome: FAO; 2013:65-66. http://www.fao.org/docrep/018/i3461e/i3461e.pdf

[5] Paige LS, Rowntree JE, Beede DK, DeLonge MS, Hamm MW. Impacts of soil carbon sequestration on life cycle greenhouse gas emissions in Midwestern USA beef finishing systems. Agricultural Systems. 2018;**162**:249-258. DOI: 10.1016/j.agsy.2018.02.003

[6] Smith P. Do grasslands act as a perpetual sink for carbon? Global Change Biology. 2014;**20**(9):2708-2711. DOI: 10.1111/gcb.12561

[7] CSIRO, Commonwealth Scientific and Industrial Research Organization. Deserts "greening" from rising CO_2. 2013. Available from: http://www.csiro.au/en/News/News-releases/2013/Deserts-greening-from-rising-CO2

[8] Brandt M, Rasmussen K, Peñuelas J, Tian F, Schurgers G, Verger A, Mertz O, Palmer JRB, Fensholt R. Human population growth offsets climate-driven increase in woody vegetation in sub-Saharan Africa. Nature Ecology & Evolution. 2017;**1**:81. DOI: 10.1038/s41559-017-0081

[9] Zhu Z, Piao S, Myneni RB, et al. Greening of the Earth and its drivers. Nature Climate Change. 2016;**6**:791-795. DOI: 10.1038/nclimate3004

[10] NASA, National Aeronautics and Space Administration. Carbon Dioxide Fertilization Greening Earth, Study Finds. 2016. Available from: http://www.nasa.gov/feature/goddard/2016/carbon-dioxide-fertilization-greening-earth

[11] Gervais F. Anthropogenic CO_2 warming challenged by 60-year cycle. Earth-Science Reviews. 2016;**155**:129-135. DOI: 10.1016/j.earscirev.2016.02.005

[12] Campbell JE, Berry JA, Seibt U, Smith SJ, Montzka SA, Launois T, Belviso S, Bopp L, Laine M. Large historical growth in global terrestrial gross primary production. Nature. 2017;**544**:84-87. https://www.nature.com/articles/nature22030

[13] Goklany LM. Carbon Dioxide: The Good News. The Global Warming Policy Foundation, Report 18. 2015. Available from: http://www.thegwpf.org/content/uploads/2015/10/benefits1.pdf

[14] Center for the Study of Carbon Dioxide and Global Change. CO_2 Science. Plant Growth Database. 2018. Available from: http://www.co2science.org/data/plant_growth/plantgrowth.php

[15] Burke M, Davis WM, Diffenbaugh NS. Large potential reduction in economic damages under UN mitigation targets. Nature. 2018;**557**:549-553. DOI: 10.1038/s41586-018-0071-9

[16] Moore P. The Positive Impact of Human CO_2 Emissions on the Survival of Life on Earth. Winnipeg, Canada: The Frontier Centre for Public Policy; 2016:1-24. https://fcpp.org/sites/default/files/documents/Moore%20-%20Positive%20Impact%20of%20Human%20CO2%20Emissions.pdf

[17] Deryng D, Elliott J, Folberth C, et al. Regional disparities in the beneficial effects of rising CO_2 concentrations on crop water productivity. Nature Climate Change. 2016. DOI: 10.1038/nclimate2995. https://www.nature.com/articles/nclimate2995

[18] Zwick S. Ruminations: Methane Math and Context. LA Chefs Column. 2018. Available from: https://lachefnet.wordpress.com/2018/05/04/ruminations-methane-math-and-context/

[19] IPCC, Intergovernmental Panel on Climate Change. Guidelines for National Greenhouse Gas Inventories. Agriculture, Forestry and Other Land Use. Vol. 4. 2016. Available from: http://www.ipcc-nggip.iges.or.jp/public/2006gl/vol4.html

[20] Nichols KL, Del Grosso SJ, Derner JD, Follett RF, Archibeque SL, Stewart CE, Paustian KH. Nitrous oxide and methane fluxes from cattle excrement on C3 pasture and C4-dominated shortgrass steppe. Agriculture Ecosystems and Environment. 2016;**225**:104-115. DOI: 10.1016/j.agee.2016.03.026

[21] Quirk TW. Twentieth century sources of methane in the atmosphere. Energy & Environment. 2010;**21**:251-265. DOI: 10.1260/0958-305X.21.3.251

[22] Glatzle A. Severe Methodological Deficiencies Associated with Claims of Domestic Livestock Driving Climate Change. Journal of Environmental Science and Engineering B. 2014:**2**:586-601. DOI: 10.17265/2162-5263/2013.10.004. http://www.davidpublisher.org/index.php/Home/Article/index?id=2117.html

[23] Glatzle A. Questioning key conclusions of FAO publications 'Livestock's Long Shadow' (2006) appearing again in 'Tackling Climate Change Through Livestock' (2013). Pastoralism: Research, Policy and Practice. 2014;**4**:1. DOI: 10.1186/2041-7136-4-1

[24] Schwietzke S, Sherwood OA, Bruhwiler LMP, Miller JB, Etiope G, Dlugokencky EJ, Michel SE, et al. Upward revision of global fossil fuel methane emissions based on isotope database. Nature. 2016;**538**:88-91. DOI: 10.1038/nature19797

[25] Schneising O, Buchwitz M, Burrows JP, Bovensmann H, Bergamaschi P, Peters W. Three years of greenhouse gas column-averaged dry air mole fractions retrieved from satellite—Part 2: Methane. Atmospheric Chemistry and Physics. 2009;**9**:443-465. https://www.atmos-chem-phys.net/9/443/2009/acp-9-443-2009.pdf and http://www.iup.uni-bremen.de/sciamachy/NIR_NADIR_WFM_DOAS/xch4_v1_2003-2005.png

[26] Wolf J, Asrar GR, West TO. Revised methane emissions factors and spatially distributed annual carbon fluxes for global livestock. Carbon Balance and Management. 2017;**12**:16. DOI: 10.1186/s13021-017-0084-y

[27] Mapfuno L, Grobler SM, Mupangwa JF, Scholtz MM, Muchenje V. Enteric methane output from selected herds of beef cattle raised under extensive arid rangelands. Pastoralism Research, Policy and Practice. 2018;**8**:15. DOI: 10.1186/s13570-018-0121-9

[28] Glatzle AF. Planet at risk from grazing animals? Tropical Grasslands—Forrajes Tropicales. 2014;**2**:60-62. DOI: 10.17138/tgft(2)60-62

[29] Garnett T, Godde C. Grazed and Confused. Food Climate Research Network. 2017. Available from: https://fcrn.org.uk/sites/default/files/project-files/fcrn_gnc_report.pdf

[30] IPCC, Intergovernmental Panel on Climate Change. The Physical Science Basis. Contribution of Working Group I to the Fourth Assessment Report. 2007. Available from: http://www.ipcc.ch/publications_and_data/ar4/wg1/en/contents.html

[31] Christy JR. U.S. House Committee on Science, Space & Technology: 2 Feb. 2016, Testimony of John R. Christy, University of Alabama in Huntsville. 2106. Available from: http://docs.house.gov/meetings/SY/SY00/20160202/104399/HHRG-114-SY00-Wstate-ChristyJ-20160202.pdf

[32] Jing-Jia L, Wang G, Dommenget D. May common model biases reduce CMIP5's ability to simulate the recent Pacific La Niña-like cooling? Climate Dynamics. 2018;**50**(3-4):1335-1351. https://link.springer.com/article/10.1007%2Fs00382-017-3688-8

[33] Hanna H, Fettweis X, Hall RJ. Recent changes in summer Greenland blocking captured by none of the CMIP5 models. The Cryosphere. 2018. DOI: 10.5194/tc-2018-91

[34] Ollila A. Challenging the scientific basis of the Paris climate agreement. International Journal of Climate Change Strategies and Management. 2018:1-17. DOI: 10.1108/IJCCSM-05-2017-0107

[35] Christy JR, Spencer RW, Braswell WD, Junod R. Examination of space-based bulk atmospheric temperatures used in climate research. International Journal of Remote Sensing. 2018;**39**(11):2580-3607. DOI: 10.1080/01431161.2018.1444293

[36] Mörner N-A. Sea level manipulation. International Journal of Engineering Science Invention. 2017;**2319**:48-51. DOI: 10.13140/RG.2.2.28591.12963. https://www.researchgate.net/publication/313854069_Sea_Level_Manipulation

[37] Patzelt G. Die nacheiszeitliche Klimaentwicklung in den Alpen im Vergleich zur Temperaturentwicklung der Gegenwart. Europäischen Institutes für Klima und Energie. Band. 2014;**3**:3-19. http://www.eike-klima-energie.eu/climategate-anzeige/buchtipp-die-nacheiszeitliche-klimaentwicklung-in-den-alpen-im-vergleich-zur-temperaturentwicklung-der-gegenwart/

[38] Maruschzik J. Klimawandel und die Gletscher in den österreichischen Alpen als Zeitzeugen. Glasshouse-Interview mit Prof. Dr. Gernot Patzelt. Glasshouse Center for Studies on a Free Economy. 2010;**1**:1-6

[39] Mangini A, Spötl C, Verdes P. Reconstruction of temperature in the Central Alps during the past 2000 yr from a $\delta^{18}O$ stalagmite record. Earth and Planetary Science Letters. 2005;**235**:741-751. DOI: 10.1016/j.epsl.2005.05.010

[40] Kullman L. Further details on holocene treeline, glacier/ice patch and climate history in Swedish Lapland. International Journal of Research in Geography. 2017;**3**(4):61-69. DOI: 10.20431/2454-8685.0304008

[41] Alley RB. The younger Dryas cold interval as viewed from Central Greenland. Quaternary Science Reviews. 2000;**19**:213-226. DOI: 10.1016/S0277-3791(99)00062-1

[42] Fudge DJ, Markle BR, Cuffey KM, et al. Variable relationship between accumulation and temperature in West Antarctica for the past 31,000 years. Geophysical Research Letters. 2016. DOI: 10.1002/2016GL068356

[43] IPCC, Intergovernmental Panel on Climate Change. Climate Change 2013. The Physical Science Basis. Contribution of Working Group I to the Fifth Assessment Report. 2013. Available from: http://www.ipcc.ch/report/ar5/wg1/

[44] Lockwood M, Owens MJ, Barnard L, Davis CJ, Steinhilber F. The persistence of solar activity indicators and the descent of the Sun into Maunder Minimum conditions. Geophysical Research Letters. 2011;**38**:22. DOI: 10.1029/2011GL049811

[45] Svensmark H, Friis-Christensen E. Variation of cosmic ray flux and global cloud coverage—A missing link in solar-climate relationships. Journal of Atmospheric and Solar-Terrestrial Physics. 1997;**59**:1225-1232. DOI: 10.1016/S1364-6826(97)00001-1

[46] Svensmark H, Enghoff MB, Shaviv NJ, Svensmark J. Increased ionization supports growth of aerosols into cloud condensation nuclei. Nature Communications. 2017;**8**:2199. DOI: 10.1038/s41467-017-02082-2

Chapter 6

Evaluation and Prediction of the Nutritive Value of Underutilised Forages as Potential Feeds for Ruminants

Mehluli Moyo, Siyabonga T. Bhiya,
Masande Katamzi and Ignatius V. Nsahlai

Additional information is available at the end of the chapter

http://dx.doi.org/10.5772/intechopen.83643

Abstract

The aim of the chapter was to evaluate and predict the nutritive and feeding value of unknown and underutilised forages. Underutilised forages were collected from various regions. Chemical composition and degradability of forages in the rumen were determined. A dataset was created bearing degradability parameters of feeds from 40 studies. Using the dataset, a step-wise regression procedure was used to develop regression equations to predict rumen degradability. Of the underutilised forages, crude protein content tended to be double for *Brassica oleracea var. acephala* compared to *Colophospermum mopane* leaves and pods. Forage grasses tended to have very low crude protein contents compared to legumes and concentrates. Underutilised *Brassica oleracea var. acephala* tended to have higher crude protein levels compared to commonly used protein sources. The regression model for predicting the soluble fraction accounted for 59% (development) and 71% (validation) of the variation. The regression model for predicting the potential degradability accounted for 65% (development) and 24% (validation) of the variation. In conclusion, the nutritive value of underutilised forages was good, high in crude protein and high potential degradability. After correcting for factors that significantly affected degradability parameters, predicted solubility and effective degradability lay near the ideal prediction line, giving good predictions.

Keywords: *Afzelia quanzensis*, *Brassica oleracea var. acephala*, *Colophospermum mopane*, degradability, feeding value

IntechOpen

1. Introduction

Ruminants such as cattle, goats and sheep are important livestock for resource-limited farmers around the world because of their ability to utilise readily available and cheap fibrous feeds that are otherwise not consumed by humans and monogastric livestock. Key to their ability to utilise feeds of high fibre content is the presence of fibrolytic bacteria in the rumen. There are a large number of plant species that have the potential of being used as forage for ruminants. Among them are a wide range of plants that are unknown to the public domain and some that are underutilised because of inadequate information on their feeding value. Exploration of these plant species is important in increasing the forage base for livestock farmers under gradually changing climatic conditions that are projected to reduce forage availability, quantity and quality. Determination of whether a forage crop can be a potential feed for a ruminant entails evaluation of its feeding value. Feeding value and quality of forages as feed for ruminants are evaluated through determining chemical composition, intake, palatability, acceptability and digestibility in vivo or in sacco. Degradability of feeds in sacco is one of the most widely used techniques to determine how much feed is digested in the rumen [1] and is important in determining feed intake. In developing countries, lack of rumen cannulated animals and/or nylon bags may hinder assessment of forage quality using rumen degradability of forages in sacco. There is a need for the development of simpler methods for the prediction of rumen degradation of forages. Simulation of digestibility of forages that has never been studied before is crucial for preliminary identification and selection of relatively unknown forages as a feed source for ruminants.

The broad objective of this chapter was to review, evaluate and predict the nutritive and feeding value of unknown and underutilised forages that have a potential of being ruminant feeds. The aim of this study was to: (1) evaluate rumen degradation of legume forages (*Colophospermum mopane* leaf meal and pods, cowpea haulms, *Mucuna pruriens,* cassava peels and *Afzelia quanzensis* legume pods), grass forages (millet stover, maize stover, maize leaves, veld grass hay and wheat straw) and *Brassica oleracea var. acephala;* and (2) predict the rumen degradation of the above-mentioned forages based on chemical composition of plant material and animal properties.

2. Review of relatively underutilised plants for feeding ruminants in sub-Saharan Africa

Non-conventional feeds and forages are feed resources used locally by farmers or have not been traditionally used in commercial or local feeding of livestock. These feeds can be available mostly with smallholder farmers and are used for short period of time, especially during the dry season when there is shortage of feeds. Literature has shown that non-conventional feeds (e.g. home waste) and forages (some forbs) are mainly used by smallholder farmers to cope during the dry season [2–4]. Although these non-convention forages are used occasionally, some of them have shown good quality attributes, which can sustain any ruminant livestock if they are provided a good quantity throughout the year. For example, bitter leaf (Vernonia), corn plant, snake weed and commelina [5] have an acceptable metabolisable energy (ME) of >7 MJ/kg DM, which is comparable to well-known Lucerne hay (7.8 MJ/kg DM; [6]). Browse plants include *Gmelina arborea, Myrianthus arboreus, Terminalia catappa, Dacroydes edulis, Parkia filicoidea* and *Tephrosia braceteolata* [7], *Moringa oleifera* (Adediran, A per com.) and accession of *Sesbania sesban*. The young leaves of *Myrianthus arboreus* (native of Angola, Cameroon, Congo, Cote d'Ivoire, Kenya, Sudan, Tanzania, Uganda and Nigeria) are popularly consumed in West Africa as vegetables and contain appreciable levels of protein, calcium, iron and phosphorous [8]. Nutrient profile of the fresh leaves of *Gmelina arborea* (originates from Southeast Asia but is planted in tropical Africa) revealed appreciable levels of crude protein (146 g/kg DM) and ether extract (127 g/kg DM) [9]. *Dacroydes edulis* can substitute 40–60% maize in poultry without any effect on production, yet it is rich in alkaloids [7]. Other energy- and protein-rich feeds are *Guizotia abyssinica* (Noug seed cake), *Hevea brasiliensis* (Rubber seed cake), *Leucaena leucocephala* leaves and pods, citrus pulp, jackfruit, palm kernel meal, tea waste, millet (seeds, bran, stover) and coconut pith. Banana leaves and pseudostems [10], cassava and cacti (high in water use efficiency, high in insoluble carbohydrates, calcium, potassium and vitamin A, but are low in crude fibre and crude protein), pineapple waste and palm oil mill effluents can be considered as a source of water for ruminants raised under harsh environments [11, 12]. Other feeds with considerable amount of water are potato peeling waste, sugar cane tops, tomato waste, apple waste, cassava peels, starch and milk waste, cocoa pods, mango seed meal and corn steep liquor.

The improvement of these feed resources could increase its availability year-round and reduce the length of the critical period when feed is in short supply. However, a cursory review of literature has depicted a paucity of information on efforts to improve and promote new options related to these feeds. Notwithstanding little is known about non-conventional feeds, it is not easy to encapsulate technological challenges on these feeds. Nonetheless, anecdotal information shows that technological challenges to include these feeds are related to (1) less interest on these feeds; many plant breeders are much more interested in food crops than forages, leading to poor testing and selection of the best-bet forages among the latter based on their agronomic aspects, (2) lack of information on these feeds at local prevailing conditions and on their potentiality (biomass production and nutrient value). Some of these underutilised forages are described below.

2.1. *Colophospermum mopane*

Mopane trees are widely distributed in the hot arid steppe areas of Southern Africa and are mainly concentrated between Southwestern Zimbabwe and Northeastern Botswana. Mopane shrubs grow in hot, dry, low-lying areas with alkaline soils. During periods of feed scarcity, cattle, goats and sheep tend to browse on Mopani tree leaves and pods. Goats prefer to browse on Mopane leaves and pods when they are reddish-brown in colour probably coinciding with high pH > 5 and low levels of condensed tannins. *Colophospermum mopane* leaves and fruits constituted 66–68% of total stomach contents of Giraffe in a low-altitude sub-tropical lowveld/bushveld mostly on the savanna habitat in winter [13]. Studies have evaluated Mopane leaf meals as a potential protein source for monogastrics, mainly in pig diets [14, 15]. Crude protein content of Mopane leaves is about 85.6 [16] and 139·6 g/kg [14]. *Colophospermum mopane* leaves had significantly lower fibre-bound proanthocyanidins (2.4 vs. 2.9 g/kg) and ytterbium-perceptible phenolic (203.8 vs. 428 g/kg) content compared to the commonly studied legume tree species such as *Acacia karroo* [14]. Few studies including Lukhele and Van Ryssen [17] and Dambe et al. [18] have evaluated the potential of *Colophospermum mopane* leaves as a feed source for ruminants, but did not determine its degradability in the rumen. This suggests that *Colophospermum mopane* forage may well be a good source of supplementary dietary protein for ruminants although more research needs to be done to increase knowledge on its feeding value for ruminants.

2.2. *Brassica oleracea var. acephala*

Commonly known as African kale, Chou Moellier and/or chomollier, this plant species thrives in well-drained soils with good soil quality and may be grown after turning in a green manure such as vetch or clover. Predominantly grown as a vegetable crop for human consumption [19, 20], little is known of the nutritional value of Chou Moellier leaves as a supplement feed source for ruminants, especially goats and sheep. There are claims that dairy cattle farmers in some parts of Australia and New Zealand use *Brassica oleracea var. acephala* leaves as a supplementary forage for dairy cows. Crude protein content of *Brassica oleracea* ranges from 15.7–25% [21, 22]. Few studies, including Barry et al. [23] and Cassida et al. [24], have evaluated the potential use of *Brassica oleracea* spp. as feed for sheep. However, the authors [25] claim that lamb growth performance (100–150 g/day) was inferior relative to the high nutritive value of *Brassica oleracea* leaves. Body weight gains of lambs grazing on *Brassica oleracea* were slightly lower than those of lambs grazing on a popular protein source, Lucerne hay (62 vs. 91 g/day) [23]. Total tract digestibility of organic matter was high for *Brassica oleracea* diets (875 g/kg) compared to Lucerne hay (731 g/kg) [23].

2.3. *Manihot esculenta*

Although the cassava root remains a good source of food for humans, cassava peels and chips may be fed to ruminants as household waste to provide supplementary nutrients. Tested in cattle, the a-fraction and effective degradability of dry matter, organic matter and crude protein were highest for cassava chips compared to generally preferred energy concentrates namely, ground corn, broken rice, rice bran and rice pollard [26]. Supplementation of rice straw with sun-dried cassava (at 1% body weight) foliage increased dry matter intake (+1341 g/d), crude protein intake (+239 g/d) and average daily gain (+201 g/d) compared to unsupplemented rice straw fed heifers [27]. In addition, molar proportions of propionic acid were higher in cattle supplemented with cassava at 2 and 3% body weight, leading to significantly low acetate: propionate ratio in the rumen [28]. Fermentation shifts towards propionic acid production are implicated in reduction in methane emissions from the rumen. The response of microbial nitrogen supply to increased levels of supplementation of cassava was a positive quadratic peaking (186.6 ± 0.85 gN/d) at 2% BW supplementation. Wanapat and Khampa [28] recommended the use of a cost-effective option to supplement using cassava at inclusion rates of 2% body weight by smallholder beef and dairy farmers. Cassava may thus play a critical role in improving the nutritional status of ruminants in tropical and sub-tropical areas coupled by its environmentally friendly role of reducing methane emissions.

2.4. *Sclerocarya birrea* ssp. *caffra*

The Marula tree fruit is a common feed supplement for ruminants in parts of Northwestern Nigeria [29], but generally not fully exploited in most parts of Southern Africa, given its abundance in the region. Full exploitation of Marula oil cake (MOC) as a supplement in ruminant diets may be limited by the scarcity of its feeding value for ruminants. Crude protein content of MOC is about 324–472 g/kg [30, 31] and may be comparable with those of commonly used protein supplements, soya bean meal (SBM) and sunflower cake (SFC) [32]. Several studies have evaluated the potential benefits of MOC as a supplement for ruminants with positive results; substitution of urea with MOC as a source of nitrogen in fattening rations had no undesirable effects on dry matter feed intake (fattening ration plus urea = 6.38 vs. fattening ration plus MOC = 6.84 kg/day) and growth rate (fattening ration plus urea = 1.62 kg/d vs. fattening ration plus MOC = 1.75 kg/d) of feedlot cattle, while a combination of equal amounts of urea and MOC in the fattening ration tended to maintain similar intakes (7.07 kg/day), but yielding better growth rates (1.82 kg/d) in feedlot cattle [33]. Potential degradability (PD) of MOC in the rumen was 723–857 g/kg for dry matter, while the PD of crude protein was 844–963 g/kg [32] in goats. Nitrogen retention was higher in goats that fed grass hay supplemented with MOC (2.8 g/d) compared to SBM (1.1 g/d) and SFC (−0.6 g/d) [32]. This suggests that *Sclerocarya birrea* ssp. *caffra* could well be a good source of supplementary dietary protein for ruminants.

2.5. *Mucuna pruriens*

With appreciable amounts of crude protein of 180–255 g/kg [34], pre-suckling kids grazing and supplemented with *Mucuna pruriens* bean had superior body weight gain (+130 vs. +86 g/day) compared to unsupplemented grazing kids, while growing lambs grazing and supplemented with *Mucuna pruriens* bean had superior body weight gain (+95 vs. +63 g/day) compared to unsupplemented grazing [35]. At similar dietary crude protein levels, *Mucuna pruriens* (inclusion level = 242 g/kg) had higher microbial protein (MP) yields (57.0 vs. 41.8 g/day) and superior microbial efficiency (70.8 vs. 51.2 g MP/kg digestible organic matter) compared to soya bean meal (inclusion level = 84.9 g/kg) [36]. Supplementation of dairy cows grazing on Napier grass with *Mucuna pruriens* increased milk yield by 32.5% compared to unsupplemented cows [37]. This suggests that *Mucuna pruriens* may well be a good source of supplementary dietary protein for all classes of ruminants.

2.6. *Strychnos* spp.

Commonly known as Monkey orange, *Strychnos* spp., fruit is indigenous to tropical and sub-tropical Africa [38]. This plant species is drought tolerant, and grows well on drained sandy soils and rocky hills [39]. Although the fruit possesses health benefits to humans, particularly children and women [40], its carbohydrate content ranges between 154 and 161 g/kg DM [41] with an average crude protein content of 128 g/kg DM [42]. The water content of the fruit ranges between 600 and 910 g/kg DM [43, 44] hence may serve as a potential water source for ruminants in arid and semi-arid regions during periods of water scarcity. There is little evidence to show that ruminants eat the Monkey orange fruit and its hard pod covering makes it an unfavourable feed for non-bipedal animals. There is limited information on the nutritional value of the Monkey orange fruit as a feed source for livestock. Given the potential of the fruit to be used as supplementary water source, evaluation of the feeding value of the fruit may render its use as a potential dual purpose feed for ruminants and other livestock.

3. Nutritive value of some underutilised forage crops

3.1. Evaluation of the nutritional value of underutilised forages and roughages

3.1.1. Materials and methods

Underutilised forage legumes and forage trees and shrubs (non-leguminous) were collected from various regions. These forages included *Colophospermum mopane* leaves and pods (Mangwe district; 20°36′57.5″S 27°45′39.7″E), and *Brassica oleracea var. acephala* (Bulawayo; 20°09′52.1″S 28°35′00.4″E) harvested in Southwestern Zimbabwe, and *Afzelia quanzensis* legume pods (Pietermaritzburg; 29°39′45.6″S 30°24′17.9″E) harvested in South Africa.

Eleven commonly used forages (10 forage grasses and 1 legume forage) were collected in KwaZulu-Natal, South Africa. These roughages included cowpea leaves and stems (*Mucuna pruriens*), maize stover, maize leaves, maize stalks (*Zea mays*), wheat straw (*Tritium aestivum*), kikuyu grass (*Pennisetum clandestinum*), weeping love grass at mature and bloom stages (*Eragrostis curvula*), bean straw, veld grass hay (Pietermaritzburg; 29°39′45.6″S 30°24′17.9″E), veld grass hay (Dundee; 28°09′17.2″S 30°12′42.8″E) and veld grass hay (Camperdown; 29°43′40.4″S 30°31′34.9″E). The forage hays were air-dried under a shade at ambient temperature and stored.

Moisture, dry matter (Method 934.01), organic matter and ash content (Method 942.05) of these forages and roughages were analysed using the procedures described by the Association of Official Analytical Chemists [45]. Nitrogen content was determined using the LECO TruSpec nitrogen analyser (LECO FP2000, LECO, Pretoria, South Africa). Crude protein content was calculated by multiplying the nitrogen content by a factor of 6.25 (crude protein = nitrogen content × 6.25). Neutral detergent fibre, acid detergent fibre and acid detergent lignin were analysed using ANKOM A220 fibre analyser (ANKOM Technology, New York, USA). Hemicellulose content was calculated as the difference between neutral detergent fibre and acid detergent fibre content (hemicellulose = neutral detergent fibre—acid detergent fibre). The cellulose and acid detergent lignin content were determined using the method of Van Soest and Wine [46].

The nylon bag technique [1] was used to determine the degradability of forages and roughages in the rumen. Dried forages were milled to pass through a 2-mm screen using a hammer mill (Scientec hammer mill 400, Lab World Pty Ltd., Johannesburg, South Africa). Approximately 4 g of each ground forage sample was weighed into ANKOM nylon bags (ANKOM Co, Fairport, New York, USA; internal dimensions: 5 × 9 cm; pore size 50 μm) and sequentially incubated (in triplicates per time interval) in the rumen for 120, 96, 72, 48, 24, 9, 6, and 3 hours using four non-lactating Jersey cows (body weight = 330 ± 19.97). The cows were fed on veld hay (*Themeda triandra*) and supplemented with 2 kg Lucerne hay per day (**Table 1**) at Ukulinga Research Farm, Pietermaritzburg, South Africa (29°39′45.6″S 30°24′17.9″E). Incubated bags were removed and washed together with the unincubated (zero hour) bags for 30 minutes (6 cycles each lasting 5 minutes) using a semi-automatic washing machine. Washed bags were oven-dried for 48 hours at 80°C and weighed.

3.1.2. Mathematical procedures

Degradability of forages was determined using dry matter loss (DML) in nylon bags. A curve for DML against incubation time was plotted and used to inspect for outliers. The model of McDonald [47] was fitted on Statistical Analysis System 9.3 (SAS Institute Inc., Cary, NC, USA) to generate degradation parameters of the forages. The model used was as follows: $Y = a + b(1{-}e^{-c(t-L)})$, where Y is the degradability at time (t), a is the intercept, b is the potentially degradable fraction, c is the rate of degradation of b and L is the lag time. Effective degradability (ED) was calculated using a predicted passage rates for each forage. The passage rate of solid was predicted using models developed by Moyo et al. [48].

3.2. Results

Of the underutilised forages, the crude protein content tended to be double as much for *Brassica oleracea var. acephala* compared to *Colophospermum mopane* leaves and pods (**Table 2**). Forage grasses (62.9 ± 34 g/kgDM) tended to have very low crude protein contents compared to legumes (137.6 ± 69) and concentrates (177 ± 39.9). Underutilised *Brassica oleracea var. acephala* (305 g/kgDM) tended to have higher crude protein levels compared to commonly used protein sources (CSC = 222 g/kgDM).

There was not much of a difference between the potential degradability of forage grasses (651 ± 111 g/kgDM), concentrates (756 ± 95.4 g/kgDM), and forage legumes, trees and shrubs (745 ± 110.2 g/kgDM) (**Tables 3–5**).

Diets	Chemical composition of feeds and diets fed to cows (g/kg DM)							
Forages	DM	OM	CP	NDF	ADF	ADL	HEM	CEL
IPR	727	922	89	745	415	-	330	-
LH	906	911	136	524	361	-	163	-
Roughages								
Diet 1 (VGH)	933	867	69	795	603	190	192	413
LH	895	564	165	487	356	77	131	279
Diet 2 (VGH + 33% LH)	920	767	101	693	521	153	172	369
Diet 3 (VGH + 50% LH)	914	716	117	641	480	134	162	346

Table 1. Chemical composition of experimental feeds and diets fed to cows during nylon bag degradability.

DM: dry matter, OM: organic matter, N: nitrogen, NDF: neutral detergent fibre, ADF: acid detergent, ADL: acid detergent lignin, HEM: hemicellulose, CEL: cellulose, VGH: veld grass hay, LH: lucerne hay.

CMLB: *Colophospermum mopane* leaves brown, CMLG: *Colophospermum mopane* leaves green CMP: *Colophospermum mopane* pods, DH: *Diheteropogon hagerupii*, ET: *Eragrostis tremula*,

Group	Chemical composition of incubated feeds (g/kg DM)							
Forage legumes, trees and shrubs	DM	OM	CP	NDF	ADF	ADL	HEM	CEL
CPH	944	935	85	617	428	102	189	328
UTCPH	627	928	141	615	422	105	193	312
GNH	932	889	109	495	386	116	110	265
CMLB	906	941	132	407	199	-	208	-
CMLG	918	931	144	470	175	-	295	-
CMP	916	946	195	477	197	-	280	-
MPL	917	942	118	607	233	-	374	-
CRP	896	953	78	489	98	-	391	-
AQLP	956	978	70	628	220	-	408	-
BOAL	908	746	305	363	137	-	226	-
Forage grasses								
MS	930	828	96	718	614	118	104	496
ML	925	660	102	645	559	100	86	459
WS	878	800	42	764	691	175	73	516
EC	931	836	107	815	503	130	312	373
ECB	925	890	128	874	615	171	259	444
KG	919	833	99	778	666	189	112	477
VGHD	932	887	41	885	629	159	256	470
VGHC	929	866	41	834	564	110	270	454
$VGHP_1$	932	877	51	849	619	189	230	430
$VGHP_2$	936	882	37	876	609	142	267	467
DH	970	959	20	880	565	78	316	485
UTDH	617	968	36	876	566	88	310	476
ET	969	976	21	796	465	67	330	397
UTET	613	971	47	829	485	72	344	414
SE	949	954	22	813	518	49	295	447
UTSE	626	956	49	812	541	90	270	450
MIS	954	913	39	816	518	130	298	394
UTMIS	619	911	46	799	523	118	276	406
SS (whole)	964	845	23	791	535	198	257	296
UTSS (whole)	624	843	45	773	514	189	259	331
SSLS	973	809	40	725	440	145	285	289
SSS	962	906	22	731	438	100	293	332
Concentrates								
MB	919	897	146	513	122	63	391	61
WB	953	951	163	477	125	35	352	86
CSC	980	948	222	570	437	101	133	339

Table 2. Chemical composition of incubated forages.

MPL: *Mucuna pruriens* leaves, MOC: marula oil cake, AQLP: *Afzelia quanzensis* legume pods, BOAL: *Brassica oleraceae var. acephala* leaves, MS: maize stover, ML: maize leaves, MT: maize stalks, MIS: millet stover, UTMIS: urea-treated millet stover, WS: wheat straw, EC: *Eragrostis*

	Rumen degradation of feeds						
	CPH	UTCPH	GNH	CMLB	CMLG	CMP	
a (g/kg)	234	236	305	519	358	398	
b (g/kg)	466	483	457	224	361	286	
c (h^{-1})	0.15	0.08	0.14	0.06	0.07	0.11	
PD (g/kg)	700	719	762	743	719	684	
ED (g/kg)	556	505	621	668	611	623	
tL (h)	-	-	-	7.1	0	1.9	
	MPL	CRP	AQLP	BOAL	CSC	MB	WB
a (g/kg)	178	293	278	351	276	449	457
b (g/kg)	550	600	274	600	371	374	342
c (h^{-1})	0.08	0.22	0.05	0.15	0.05	0.29	0.22
PD (g/kg)	728	893	552	951	647	823	799
ED (g/kg)	578	821	449	851	439	755	722
tL (h)	0	0	2.0	0	-	-	-

Table 3. Nylon bag degradation of forage legumes, forage trees and shrubs (non-leguminous), and concentrates. ED was calculated at kp: rate of passage of particles in the rumen = 0.03 per h.

Diet 1 (100% veld hay)										
	MS	ML	WS	ECM	ECB	KG	GHD	GHC	GHP_1	GHP_2
a (g/kg)	194	158	17	86	43	76	53	44	39	174
b (g/kg)	445	454	373	518	491	430	475	400	446	439
c (h^{-1})	0.049	0.049	0.033	0.048	0.037	0.047	0.027	0.032	0.026	0.029
PD (g/kg)	639	612	391	604	534	506	499	445	486	613
ED (g/kg)	645	637	351	622	454	339	405	398	385	351
Diet 2 (67% veld hay: 33% lucerne hay)										
a (g/kg)	194	158	16	87	44	76	24	43	51	28
b (g/kg)	592	623	534	726	818	577	869	622	521	647
c (h^{-1})	0.039	0.043	0.016	0.033	0.017	0.017	0.008	0.018	0.029	0.017
PD (g/kg)	786	780	556	813	819	652	894	665	572	676
ED (g/kg)	531	524	201	467	289	272	212	258	285	265
Diet 2 (50% veld hay: 50% lucerne hay)										
a (g/kg)	194	158	16	87	44	76	24	43	40	173
b (g/kg)	607	659	489	727	647	507	593	591	543	428
c (h^{-1})	0.052	0.051	0.033	0.042	0.029	0.03	0.019	0.024	0.025	0.03
PD (g/kg)	801	817	505	814	691	582	616	634	583	600
ED (g/kg)	579	572	269	512	365	329	256	299	285	387

Table 4. Nylon bag degradability of forage grasses (roughages) in cows fed with three different diets. ED was calculated at kp: rate of passage of particles in the rumen = 0.03 per h.

	Rumen degradation of feeds					
	DH	UTDH	ET	UTET	SE	UTSE
a (g/kg)	99	129	136	165	91	157
b (g/kg)	572	529	521	538	564	525
c (h^{-1})	0.02	0.02	0.02	0.02	0.02	0.03
PD (g/kg)	671	658	657	703	655	682
ED (g/kg)	224	248	267	271	238	307
	MS	UTMS	SS (whole)	UTSS (whole)	SSLS	SSS
a (g/kg)	131	182	223	251	217	206
b (g/kg)	552	437	507	470	548	421
c (h^{-1})	0.01	0.02	0.02	0.02	0.03	0.02
PD (g/kg)	683	619	730	721	765	627
ED (g/kg)	214	298	330	362	376	313

Table 5. Nylon bag degradability of urea treated and untreated forage grasses (roughages) in cows fed kikuyu pasture.

curvula, ECB: *Eragrostis curvula* at bloom stage, KG: kikuyu grass, SE: *Schizachyrium exile*, VGHD: veld grass hay from Dundee, VGHC: veld grass hay Camperdown, $VGHP_1$: veld grass hay Pietermaritzburg area 1, $VGHP_2$: veld grass hay from the Pietermaritzburg area 2, CPH: cowpea husks, CRP: cassava root peels, GNH: groundnut haulms, UTCPH: urea-treated cowpea husks, UTDH: urea-treated *Diheteropogon hagerupii*, UTET: urea-treated *Eragrostis tremula*, UTSE: urea-treated *Schizachyrium exile*, UTMIS: urea-treated maize stover, SS: sorghum stover, UTSS: urea-treated sorghum stover, SSLS: sorghum stover leaves and sheath, SSS: sorghum stover stems, MB: millet bran, WB: wheat bran, and CSC: cottonseed cake.

CMLB: *Colophospermum mopane* leaves—brown, CMLG: *Colophospermum mopane* leaves - green, CMPG: *Colophospermum mopane* pods, CPH: cowpea husks, CRP: cassava root peels, GNH: groundnut haulms, MPL: *Mucuna pruriens* leaves, AQLP: *Afzelia quanzensis* legume pods, BOAL: *Brassica oleraceae var. acephala* leaves, UTCPH: urea-treated cowpea husks, MB: millet bran, WB: wheat bran, CSC: cottonseed cake, a: rapidly degradable fraction, b: slowly degradable fraction, c: rate of degradation, PD: potential degradability, and ED: effective degradability.

MS: maize stover, ML: maize leaves, MT: maize stalks, WS: wheat straw, EC: *Eragrostis curvula*, ECB: *Eragrostis curvula* at bloom stage, KG: kikuyu grass, VGHD: veld grass hay from Dundee, VGHC: veld grass hay Camperdown, VGHP1: veld grass hay Pietermaritzburg area 1, VGHP2: veld grass hay from the Pietermaritzburg area 2, kp: rate of passage of particles in the rumen, a: rapidly degradable fraction, b: slowly degradable fraction, c: rate of degradation, PD: potential degradability, and ED: effective degradability.

MS: maize stover, ML: maize leaves, MT: maize stalks, WS: wheat straw, EC: *Eragrostis curvula*, ECB: *Eragrostis curvula* at bloom stage, KG: kikuyu grass, VGHD: veld grass hay.

4. Is it possible to predict the rumen digestibility (feeding value) of unknown and underutilised forages?

4.1. Prediction of degradation of forages in the rumen using feed and animal properties

4.1.1. Materials and methods

Data were collected from studies that reported at least average values for in sacco (nylon bag technique) degradability parameters (a, soluble fraction; b, slowly degradable fraction and c, rate of degradation) of roughages and stated the diet, feeds and feed supplements given to animals. A dataset was created bearing degradability parameters from wild and domesticated ruminants from 40 studies. Factors affecting degradability were identified in each of these studies and were categorised into two main groups: (1) diet properties (i.e. fed to the animal) and (2) feed sample properties (i.e. incubated in the rumen). Diet properties were used to account for the effects of rumen ecology on fermentation and included neutral detergent fibre (NDF), starch (STA) and crude protein (CP) contents of entire diet (all in g/kg), level of concentrate supplementation (%) and provision of a urea supplement in the form of a lick (presence = 1, absence = 0). Feed sample properties included urea treatment (%) of sample and feed compositional attributes (DM, dry matter; CP, crude protein; NDF, neutral detergent fibre, ADF, acid detergent fibre; HEM, hemicellulose and ash all in g/kg). Starch content of the diet fed to animals was calculated using the formula: STA = 1000–(NDF + CP). Potential degradability (PD) and hemicellulose (HEM) content were calculated in studies that did not report them using the formulae: PD = a + b; and HEM = NDF—ADF, respectively. Studies that did not report dietary composition of feeds but mentioned names of feeds used had their composition looked up in studies that reported them. These factors were used as input parameters to develop regression models for predicting degradability of feeds in the rumen.

A step-wise regression procedure on the Statistical Analysis System 9.3 (SAS Institute Inc., Cary, NC, USA) was used to select parameters that qualified to develop regression equations to predict (1) rapidly degradable fraction of fibre (a), (2) potential degradability (PD), (3) time lag for fermentation to occur (tL), and (4) rate of degradation (c) in the rumen. One parameter from a pair of correlated parameters was dropped in model development when both correlated parameters significantly influence degradation parameters. Those parameters that qualified for model development were CP and NDF content of feed sample (model for soluble fraction of fibre); ADF content of feed sample and STA content of diet (model for potential degradability); ADF, CP and ash content of feed sample, and STA content of diet (model for time-lag); NDF and CP content of feed sample, and, STA and DNDF content of diet (model for degradation rate).

Regression models were used to simulate the rumen degradability of *Colophospermum mopane* leaves and pods, *Diheteropogon hagerupii*, *Eragrostis tremula*, *Mucuna pruriens* leaves, Marula oil cake, *Afzelia quanzensis* legume pods, *Brassica oleraceae var. acephala* leaves, maize stover, leaves and stalks, millet stover, wheat straw, *Eragrostis curvula*, Kikuyu grass, *Schizachyrium exile*, veld grass hay, cowpea husks, cassava root peels, groundnut haulms, *Eragrostis tremula*, sorghum stover, leaves and sheath, and stems, millet bran, wheat bran, and cottonseed cake. The effective degradability of these forages was calculated using the model of McDonald [47].

4.1.2. Statistical analyses

For all evaluations, regression analyses of observed against predicted degradability were carried out using the linear regression procedure. Coefficients of determination (R^2) were used to evaluate the precision of regression lines in approximating real data points of models and standard error of the mean (SEM) was used to determine the accuracy of prediction.

4.2. Results

4.2.1. Model development

From the step-wise regression procedure for all prediction models, level of concentrate supplementation, provision of a urea supplement in the form of a lick and urea treatment of feed sample were rejected in model development.

The regression model for predicting the soluble fraction (a) was a = 558.12(±62.45) + 0.27 (±0.133) CP–0.57(±0.07) NDF (n = 113, SEM = 6.86), accounting for 59% of the variation in development.

The regression model for predicting the potential degradability (PD) was PD = 1025.96(±66.64) –0.91(±0.10) ADF + 0.32(±0.08) STA (n = 113, SEM = 9.27), accounting for 65% of the variation in development.

The regression model for predicting the time-lag (tL) was tL = −11.33(±1.89) + 0.030(±0.002) ADF + 0.01(±0.003) CP–0.006(±0.001) STA + 0.02(±0.007) ASH (n = 113, SEM = 0.17), accounting for 77% of the variation in development.

The regression model for predicting the rate of degradation (c) was c = 0.12(±0.05) + 0.00013 (±0.00002) CP–0.00012(±0.00006) STA–0.00002(±0.00001) NDF–0.00008(±0.00005) DNDF (n = 113, SEM = 0.0009), accounting for 55% of the variation in development.

4.2.2. Model predictions

The regression model for predicting the soluble fraction of feeds accounted for 70% of the variation in prediction for forage legumes, trees and shrubs, forage grasses and concentrates (**Figure 1**).

The regression model for predicting the potential degradability accounted for 24% of the variation in prediction for forage legumes, trees and shrubs, forage grasses and concentrates (**Figure 2**).

The regression model for predicting the slowly degradable fraction of feeds for forage legumes, trees and shrubs, forage grasses and concentrates (**Figure 3**).

The regression model for predicting the rate of degradation accounted for 4% of the variation in prediction for forage legumes, trees and shrubs, forage grasses and concentrates (**Figure 4**).

http://dx.doi.org/10.5772/intechopen.83643

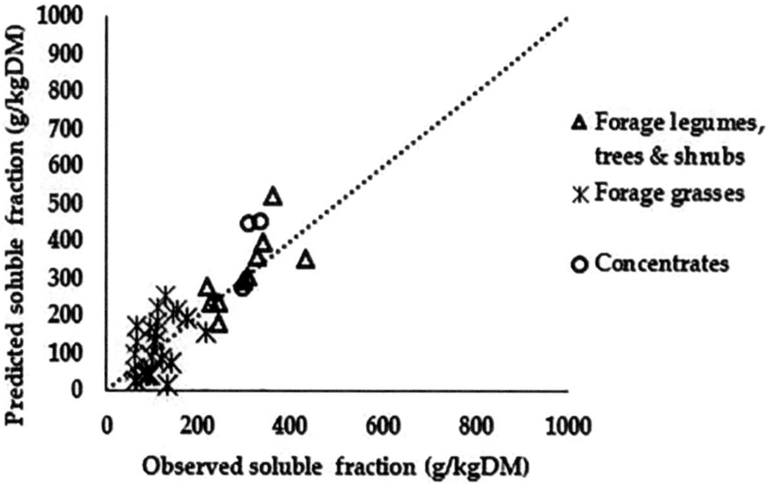

Figure 1. Relationship between observed and predicted degradability of soluble fraction.

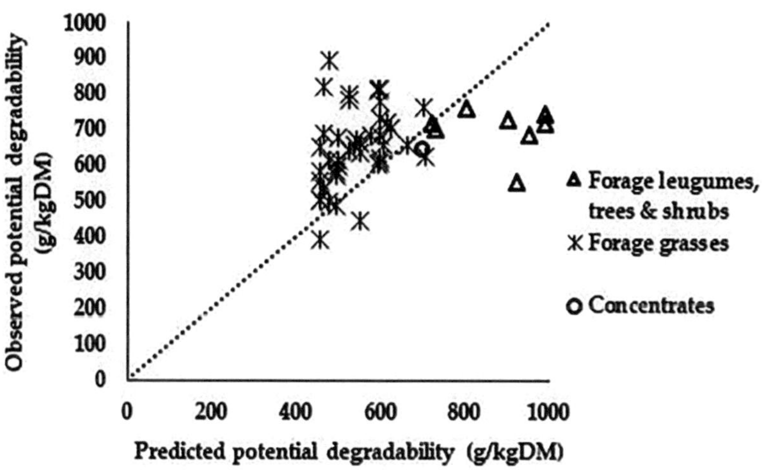

Figure 2. Relationship between observed and predicted potential degradability.

The regression model for predicting the effective degradability of feeds accounted for 57% of the variation in prediction for forage legumes, trees and shrubs, forage grasses and concentrates (**Figure 5**).

4.3. Discussion

Among the forage legumes, trees and shrubs, *Brassica oleracea var. acephala* leaves had a superior crude protein content and the lowest neutral and acid detergent fibre contents. The CP content of *Brassica oleracea var. acephala* is slightly higher than those reported by McDonald et al. [21] and Barry et al. [22]. The rate of degradation of *Colophospermum mopane* pods was similar to that of *Brassica oleracea var. acephala*. High levels of degradability of these feeds were partly due to

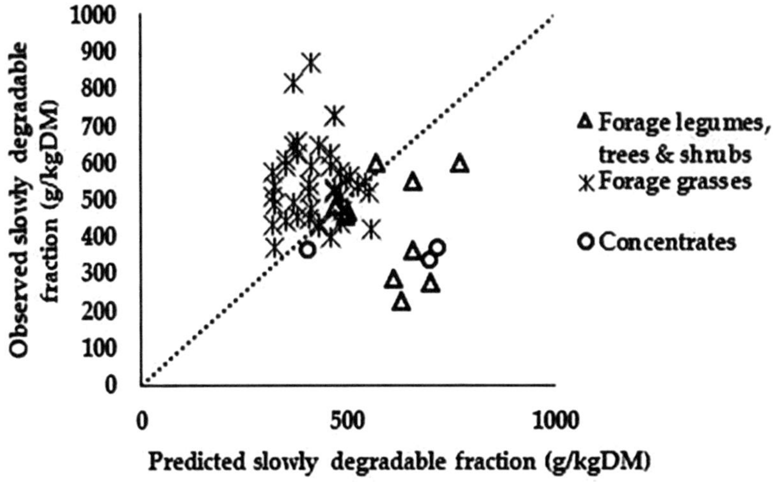

Figure 3. Relationship between observed and predicted degradability of slowly degradable fraction.

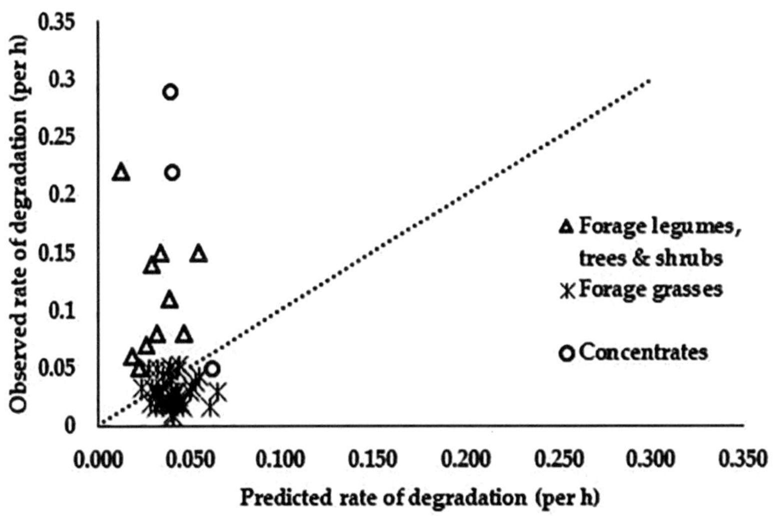

Figure 4. Relationship between observed and predicted rates of degradation .

high levels of crude protein, which could help in the proliferation of microbial populations in the rumen, increasing ED and rate of degradation of these forages. Faster rates of degradation may suggest faster rates of passage of these feeds in the rumen, which could increase microbial protein supply for host animals in the hindgut, improving animal's nutritional status. The CP level in *Colophospermum mopane* leaves was comparable to results of Halimani et al. [14], while NDF contents tended to be comparably higher than those reported by other authors [14, 17].

Compared to concentrates used in the study, *Brassica oleracea var. acephala* leaves tended to have superior crude protein levels than the 'brans' and cotton seed cake. Despite this trend, the brans tended to have faster degradation rates than cotton seed cake and *Brassica oleracea var. acephala* leaves. *Colophospermum mopane* leaves and pods had comparable CP and NDF levels compared to maize and wheat brans, suggesting that *Brassica oleracea var. acephala* and, *Colophospermum mopane* can be used as good sources of supplementary protein to ruminants.

Relationships between two variables are said to be ideal when the coefficient of determination (R^2) is in unity; any deviation from the unity degree indicates the degree of imperfection. The above parameters were used to determine the effective degradability (ED): (ED = a + (PD−a) × c/(c + kp); where 'a' is a soluble fraction, PD is the potential degradability, 'c' is the rate of degradation and kp is the rate of passage of particles through the rumen. Effective degradability is equivalent to digestibility in the rumen. The predicted effective degradability indicated in **Figure 5** followed the expected trends, suggesting that these models (for predicting 'a', PD, and 'c') in the meantime can be used for this purpose. The overall trend between the observed and the predicted digestibility is positive, though accounting for just 36–52% of the total variation [49], which does not compare favourably with R^2 of 70% obtained with the application of the simulation model to temperate roughages [43] and those from this study. The amount of variation accounted for in observed against predicted digestibility for simulations by Nsahlai and Apaloo [49, 50] was comparably higher than those reported in empirical studies by Shem et al. [10], Kibon and Orskov [51] and Umunna et al. [52].

The rather low precision in predicting the rate of degradation (mainly for concentrates, legume forages, trees and shrubs) and the potential degradability (concentrates) of feeds in this study may have been due to the fact that the studies that were used in model development reported data on degradation of roughages grasses only, which are generally of low quality, and did not use data on concentrates, legume forages, trees and shrubs. Despite this, simulations of solubility and effective degradability were good, suggesting that slight modification of model parameters may give better prediction of all degradability (nutritive value) of a large number and classes of forage crops. Generally, there is a poor simulation of

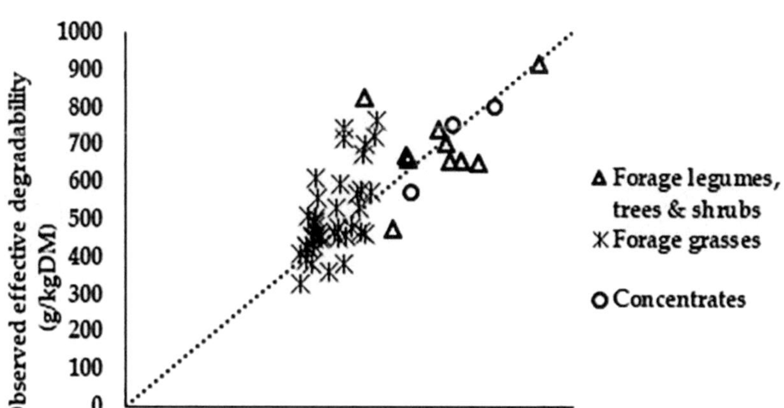

Figure 5. Relationship between observed and predicted effective degradability.

digestibility for low quality roughages, which are commonly grazed and fed to ruminants in the tropics. Ambient temperature grossly affects the digestibility of plant material through its influence on lignin deposition in plants. Studies should focus on development of digestibility models that account for variability in diet quality as brought about by ambient temperature. Future studies may need to account for the type of model used in computation of degradation parameters.

5. Conclusions

The nutritive value of underutilised forages, *Brassica oleracea var. acephala* and, *Ṅ* leaf meal and pods was good with high levels of crude protein and potential degradability in the rumen, suggesting their potential use as ruminant feeds during the dry season. Predicted solubility and effective degradability lay near the ideal prediction line, giving good predictions for these parameters. However, some adjustments in the inputs for prediction of potential degradability and rate of degradation are needed to improve predictions.

Acknowledgements

This study was financially supported by the National Research Foundation (NRF) of the Republic of South Africa (project name: Modelling of intake, feeding behaviour and kinetics of digestion and passage of digesta in ruminants; grant unique number: 112905) and the University of KwaZulu-Natal (competitive grant number: P029).

Conflict of interest and declaration

The authors declare that they have no competing interests. We affirm that all the authors of this manuscript agree to the submission, and the manuscript has not been submitted to be published in or considered for publication anywhere else. The views expressed in the paper are those of the authors and not of the National Research Foundation (NRF) of the Republic of South Africa.

Author details

Mehluli Moyo, Siyabonga T. Bhiya, Masande Katamzi and Ignatius V. Nsahlai*

*Address all correspondence to: nsahlaii@ukzn.ac.za

Animal and Poultry Science, School of Agricultural, Earth and Environmental Sciences, University of KwaZulu-Natal, Pietermaritzburg, South Africa

References

[1] Orskov ER, DeB Hovell FD, Mould F. The use of the nylon bag technique for the evaluation of feedstuffs. Tropical Animal Health and Production. 1980;**5**:195-213

[2] Delve RJ, Cadisch G, Tanner JC, Thorpe W, Thorne PJ, Giller KE. Implications of livestock feeding management on soil fertility in the smallholder farming systems of sub-Saharan Africa. Agriculture, Ecosystems and Environment. 2001;**84**:227-243

[3] Njarui DMG, Gatheru M, Wambua JM, Nguluu SN, Mwangi DM, Keya GA. Feeding management for dairy cattle in smallholder farming systems of semi-arid tropical Kenya. Livestock Research for Rural Development. 2011;**23**, Article #111. Retrieved July 30, 2013

[4] Mutimura M, Everson TM. Assessment of livestock feed resource-use patterns in low rainfall and aluminium toxicity prone areas of Rwanda. African Journal of Agricultural Research. 2011;**6**(15):3461-3469

[5] Mutimura M, Ebong C, Rao IM, Nsahlai IV. Nutritional values of available ruminant feed resources in smallholder dairy farms in Rwanda. Tropical Animal Health and Production. 2015;**47**:1131-1137

[6] McDonald P, Edwards RA, Greenhalgh JFD, Morgan CA, Sinclair LA, Wilkinson RG. Animal Nutrition. 7th ed. Harlow, England: Pearson Education Limited; 2011. 692 p

[7] Amata IA. The use of non-conventional feed resources (NCFR) for livestock feeding in the tropics: A review. Journal of Global Biosciences. 2014;**3**(2):604-613

[8] Okafor JC. Myrianthus arboreus. P. Beauv. In: Grubben GJH, Denton OA, editors. PROTA 2: Vegetables/Legumes. Wageningen, Netherlands: PROTA Foundation; 2004

[9] Amata IA, Lebari T. Comparative evaluation of the nutrient profile of four selected browse plants in the tropics, recommended for use as non-conventional livestock feeding materials. African Journal of Biotechnology. 2011;**10**(64):14230-14233

[10] Shem MN, Orskov ER, Kimambo AE. Prediction of voluntary dry matter intake, digestible dry-matter intake and growth of cattle from the degradation characteristics of tropical foods. Animal Science. 1995;**60**:65-74

[11] Nefzaoui A, Salem HB. Forage fodder and animal nutrition. In: Nobel PS, editor. Cacti: Biology and Uses. Berkeley, CA, USA: University of California Press; 2002. pp. 190-210

[12] Salem HB, Smith T. Feeding strategies to increase small ruminant production in dry environments. Small Ruminant Research. 2008;**77**(2):174-194

[13] Hall-Martin AJ. Food selection by Transvaal lowveld giraffe as determined by analysis of stomach contents. South African Journal of Wildlife Research. 1974;**4**(3):191-202

[14] Halimani TE, Ndlovu LR, Dzama K, Chimonyo M, Miller BG. Metabolic response of pigs supplemented with incremental levels of leguminous *Acacia karroo*, *Acacia nilotica* and *Colophospermum mopane* leaf meals. Animal Science. 2005;**81**(1):39-45

[15] Halimani TE, Ndlovu LR, Dzama K, Chimonyo M, Miller BG. Growth performance of pigs fed on diets containing *Acacia karroo, Acacia nilotica* and *Colophospermum mopane* leaf meals. Chemical Analysis. 2007;**100**(100.0):100-100

[16] Ferwerda JG. Charting the quality of forage: measuring and mapping the variation of chemical components in foliage with hyperspectral remote sensing. Wageningen, Wageningen University, 2005. ITC Dissertation 126; 2005. 166 p. ISBN 90-8504-209-7

[17] Lukhele MS, Van Ryssen JBJ. The chemical composition and potential nutritive value of the foliage of four subtropical tree species in southern Africa for ruminants. South African Journal of Animal Science. 2003;**33**(2):132-141

[18] Dambe LM, Mogotsi K, Odubeng M, Kgosikoma OE. Nutritive value of some important indigenous livestock browse species in semi-arid mixed *Mopane bushveld*, Botswana. Livestock Research for Rural Development. 2015;**27**(10)

[19] Sowing New Seeds Project, 2018. Growing African Kale (leaflet). Garden Organic (Henry Doubleday Research Association), Coventry, Warwickshire, United Kingdom. https://www.gardenorganic.org.uk/sites/www.gardenorganic.org.uk/files/sns/factsheets/FactsheetAfricanKale.pdf [Accessed 2018/06/02]

[20] Emebu PK, Anyika JU. Proximate and mineral composition of kale (*Brassica oleracea*) grown in Delta State. Nigeria. Pakistan Journal of Nutrition. 2011;**10**(2):190-194

[21] McDonald RC, Manley TR, Barry TN, Forss DA, Sinclair AG. Nutritional evaluation of kale (*Brassica oleracea*) diets: 3. Changes in plant composition induced by soil fertility practices, with special reference to SMCO and glucosinolate concentrations. The Journal of Agricultural Science. 1981;**97**(1):13-23

[22] Barry TN, Manley TR, Duncan SJ. Quantitative digestion by sheep of carbohydrates, nitrogen and S-methyl-L-cysteine sulphoxide in diets of fresh kale (*Brassica oleracea*). The Journal of Agricultural Science. 1984;**102**(2):479-486

[23] Barry TN, Manley TR, Millar KR. Nutritional evaluation of kale (*Brassica oleracea*) diets: 4. Responses to supplementation with synthetic S-methyl-L-cysteine sulphoxide (SMCO). The Journal of Agricultural Science. 1982;**99**(1):1-12

[24] Cassida KA, Barton BA, Hough RL, Wiedenhoeft MH, Guillard K. Feed intake and apparent digestibility of hay-supplemented brassica diets for lambs. Journal of Animal Science. 1994;**72**(6):1623-1629

[25] Nicol AM, Barry TN. The feeding of forage crops. In: Drew KR, Fennessy PF, editors. New Zealand Society of Animal Production Occasional Publication No. 7 Supplementary Feeding. Mosgiel New Zealand: C/-Invermay Research Centre. 1980. pp. 69-102

[26] Chumpawadee S, Sommart K, Vongpralub T, Pattarajinda V. In sacco degradation characteristics of energy feed sources in Brahman-Thai native crossbred steers. Journal of Agricultural Technology. 2005;**1**(2):192-206

[27] Sath K, Borin K, Preston TR. Effect of levels of sun-dried cassava foliage on growth performance of cattle fed rice straw. Livestock Research for Rural Development. 2008; Retrieved August 23, 2018. http://www.lrrd.org/lrrd20/supplement/sath2.htm

[28] Wanapat M, Khampa S. Effect of levels of supplementation of concentrate containing high levels of cassava chip on rumen ecology, microbial-N supply and digestibility of nutrients in beef cattle. Asian-Australasian Journal of Animal Sciences. 2007;**20**(1):75-81

[29] Muhammad N, Omogbai IJ, Maigandi SA, Abubakar IA, Shamaki SB. Quantification of *Sclerocarya birrea* (Marula) fruits as feed supplement for ruminants in dry sub-humid zone of Nigeria. World Scientific News. 2016;**46**:88-99

[30] Mdziniso PM, Dlamini AM, Khumalo GZ, Mupangwa JF. Nutritional evaluation of Marula (*Sclerocarya birrea*) seed cake as a protein supplement in dairy meal. Journal of Applied Life Sciences International. 2016;**4**(3):1-11

[31] Malebana IM, Nkosi BD, Erlwanger KH, Chivandi E. A comparison of the proximate, fibre, mineral content, amino acid and the fatty acid profile of Marula (*Sclerocarya birrea* caffra) nut and soyabean (*Glycine max*) meals. Journal of the Science of Food and Agriculture. 2018;**98**(4):1381-1387

[32] Mlambo V, Dlamini BJ, Nkambule MT, Mhazo N, Sikosana JLN. Nutritional evaluation of Marula (*Sclerocarya birrea*) seed cake as a protein supplement for goats fed grass hay. Tropical Agriculture. 2011;**41**(3216):010035-010009

[33] Mlambo V, B D, Ngwenya MD, Mhazo N, Beyene ST, Sikosana JLN. In sacco and in vivo evaluation of Marula (*Sclerocarya birrea*) seed cake as a protein source in commercial cattle fattening diets. Livestock Research for Rural Development. 2011;**23**(5)

[34] Muchadeyi R. Herbage yields, chemical composition and in-vitro digestibility of dual-purpose legumes intercropped with maize for dry season fodder supplementation. M.Sc. Thesis. Harare: University of Zimbabwe; 1998

[35] Castillo-Caamal JB, Jimenez-Osornio JJ, Lopez-Perez A, Aguilar-Cordero W, Castillo-Caamal AM. Feeding Mucuna beans to small ruminants of Mayan farmers in the Yucatan peninsula. Mexico. Tropical and Subtropical Agroecosystems. 2003;**1**(2-3)

[36] Chikagwa-Malunga SK, Adesogan AT, Szabo NJ, Littell RC, Phatak SC, Kim SC, et al. Nutritional characterization of Mucuna pruriens: 3. Effect of replacing soybean meal with Mucuna on intake, digestibility, N balance and microbial protein synthesis in sheep. Animal Feed Science and Technology. 2009;**148**(2-4):107-123

[37] Juma HK, Abdulrazak SA, Muinga RW, Ambula MK. Evaluation of Clitoria, Gliricidia and Mucuna as nitrogen supplements to Napier grass basal diet in relation to the performance of lactating Jersey cows. Livestock Science. 2006;**103**(1-2):23-29

[38] Bisset NG. The African species of Strychnos, Part I. The ethnobotany. Lloydia, 33; 1970. pp. 201-243

[39] Orwa C, Mutua A, Kindt R, Jamnadass R, Simons A. Agroforestree Database: A Tree Species Reference and Selection Guide Version 4.0. Nairobi, KE: World Agroforestry Centre ICRAF; 2009

[40] Amarteifo JO, Mosase MO. The chemical composition of selected indigenous fruits of Botswana. Journal of Applied Sciences and Environmental Management. 2006;**10**(2)

[41] Saka JK, Msonthi JD. Nutritional value of edible fruits of indigenous wild trees in Malawi. Forest Ecology and Management. 1994;**64**(2-3):245-248

[42] Mwamba CK. Monkey orange: Strychnos cocculoides (No. 8). Crops for the Future; 2006

[43] Malaisse F, Parent G. Edible wild vegetable products in the Zambezian woodland area: A nutritional and ecological approach. Ecology of Food and Nutrition. 1985;**18**(1):43-82

[44] Bello MO, Falade OS, Adewusi SRA, Olawore NO. Studies on the chemical compositions and anti-nutrients of some lesser known Nigeria fruits. African Journal of Biotechnology. 2008;**7**(21)

[45] AOAC. Official Methods of Analysis. 17th ed. Arlington VA, USA: Association of Official Analytical Chemists (AOAC); 2000

[46] Van Soest PJ, Wine RH. Determination of lignin and cellulose in acid-detergent fiber with permanganate. Journal of the Association of Official Analytical Chemists. 1968;**51**:780-785

[47] McDonald I. A revised model for the estimation of protein degradability in the rumen. Journal of Agricultural Science. 1981;**96**:251-252

[48] Moyo M, Gueguim Kana EB, Nsahlai IV. Modelling of digesta passage rates in grazingand browsing domestic and wild ruminant herbivores. South African Journal of Animal Science

[49] Nsahlai IV, Apaloo J. On the suitability of the Illius and Gordon's model for simulating the intake and digestibility of roughage diets by ruminants. South African Journal of Animal Science. 2007;**37**(4):275-289, 2017;**47**(3):362-377

[50] Illius AW, Gordon IJ. Prediction of intake and digestion in ruminants by a model of rumen kinetics integrating animal size and plant characteristics. The Journal of Agricultural Science. 1991;**116**(1):145-157

[51] Kibon A, Orskov ER. The use of degradation characteristics of browse plants to predict intake and digestibility by goats. Animal Production. 1993;**57**:247-251

[52] Umunna NN, Nsahlai IV, Osuji PO. Degradability of forage protein supplements and their effects on the kinetics of digestion and passage. Small Ruminant Research. 1995;**17**(2):145-152

Chapter 7

Ensiling Alfalfa (*Medicago sativa* L.) and Orchard Grass (*Dactylis glomerata* L.) Forage Harvested at 08:00 or 14:00, without Wilting or 1 or 2 h Wilting and with or without Use of Bacterial Inoculant

Ricardo D. Améndola-Massiotti,
Renato González-Ortiz, Luis A. Miranda-Romero,
Juan A. Burgueño-Ferreira and Pedro Topete-Pelayo

Additional information is available at the end of the chapter

http://dx.doi.org/10.5772/intechopen.81509

Abstract

Alfalfa forage is difficult to ensile due to low water-soluble carbohydrate content and high buffering capacity. The objective was to assess at Chapingo, Mexico, during the rainy season effects of combinations of harvest hours (08:00, 14:00), wilting time (0, 1, 2 h) and bacterial inoculants on the quality of silage made of alfalfa and orchard grass forage, made in 200-L containers. The experiment was conducted in three phases with two replicates per phase. Variables measured in freshly cut forage and silages were dry matter content (DM), buffer capacity, pH, and alcohol soluble carbohydrates (ASC). Silos remained sealed during 60 d, and additional variables measured in silage were aerobic stability, NH3 -N and in vitro disappearance of DM. In forage harvested at 14:00 h, DM and ASC contents were higher; pH and buffering capacity were not affected by harvest hour; in silages made of that forage, NH3-N levels were lower, while ASC contents and in vitro disappearance of MS were unaffected by harvest hour. Treatments with inoculants were less aerobic stable for 5 days when made of forage harvested at 08:00 h but more stable when made of forage harvested at 14:00 h. Harvesting at 14:00 h was advantageous as silage presented higher DM and ASC contents.

Keywords: silage, *Medicago sativa*, wilting, inoculant, harvesting time

1. Introduction

Alfalfa (*Medicago sativa* L.) grasslands are dominant in irrigated temperate Mexico; however, their yield in the dry autumn-winter season is 50% lower than during the rainy spring-summer [1]. Conservation of forage surpluses is therefore necessary; but, due to summer rains, haymaking is risky and hence silage making becomes the best option. Nonetheless, ensiling alfalfa forage is difficult because of its low content of water-soluble carbohydrates and high buffering capacity, which delay the lactic acid production and therefore hinder a rapid decline of pH within the ensiled forage [2].

Technical alternatives for ensiling alfalfa forage are to harvest in afternoon hours [3], wilting [4] and the use of lactic acid bacteria (LAB) homo- and heterofermentative inoculants with enzymes [5].

Cutting in the afternoon and wilting increased dry matter (DM) and alcohol soluble carbohydrates (ASC) concentration in alfalfa, which led to silage with improved conservation attributes such as lower pH, greater concentrations of lactate, lower concentrations of volatile fatty acids and NH3–N [3]. However, the effect of wilting largely depends on weather conditions, since events of rain during wilting, together with poor drying conditions (low potential evapotranspiration) may lead to considerable rise in pH and losses of sugars, causing worse silage fermentation [4].

The use of inoculants consisting of homofermentative bacteria hastened the drop of pH during the fermentation but did not improve the aerobic stability of the silage [6]. Such result is consistent with [7], who reported that inoculants based on homofermentative lactic acid bacteria did not improve aerobic stability in about two thirds of the cases; on the contrary, inoculant based on the heterolactic acid bacterium *Lactobacillus buchneri* increases the aerobic stability of silages mainly due to a rise in the concentration of acetic acid.

Aerobic stability is defined as the length of time that silage remains cool and does not spoil after it is exposed to air [7]. Since silage making relies on keeping anaerobic conditions, once the silo is open in the feeding phase and hence the silage is exposed to oxygen, it becomes liable to oxidation by the coming into action of dormant aerobic bacteria, yeasts and molds, which will be producing CO_2 [8], leading to rises in temperature and pH [7].

The fermentation and storage phases are dominated by anaerobic processes; as a result, silage contains yeasts, molds and some aerobic bacteria which are dormant under these anaerobic conditions. Introduction of oxygen, by deterioration of the sealing or opening the silo for the feeding phase, activates these aerobic microorganisms whose respiration consumes valuable nutrients, producing carbon dioxide and water with loss of dry matter and nutrients and decay of the silage [7].

Based on the above stated, this study was aimed at assessing during the rainy season, in three different months (different short-term weather conditions) the effect of hour at harvest, wilting and use of a mixed inoculant on quality properties of freshly cut forage from a mixed alfalfa and orchard grass (*Dactylis glomerata* L.) grassland dominated by alfalfa, and the corresponding silages.

2. Material and methods

The field work was carried out in three silage phases between June and September 2011 at Chapingo University, State of Mexico, Mexico, 19° 29 'N, 98° 54 'W and 2240 meters above sea level, under temperate sub-humid climate with summer rains.

Forage from a 2-year old alfalfa-orchard grass mixed grassland (0.37ha) was used, harvested after 42 days of regrowth, yielding on average 2480 kg DM/ha per harvest. Such mixed grasslands are regularly used within an irrigated grassland and forage crops rotation for dairy production under grazing; during the first 2 years of these grasslands, alfalfa is dominant (more than 70% of dry matter of harvested herbage). As silos, plastic containers with a capacity of 200 L, 52 cm in diameter and 95 cm in height were used, with lid and strap for air tight sealing.

The experiment comprised three phases (phase 1: June–July, phase 2: August, phase 3: September) in which 12 treatments were evaluated with two replicates in each phase. The treatments resulted from the combination of 2 × 2 × 3 complete factorial arrangement of two cutting schedules (08:00 and 14:00 h), two levels of LAB inoculant (0 and 5 g/t forage) and three wilting times (0, 1 and 2 h). During each day of operation, two containers were filled (one with forage cut at 08:00 h and the other with forage cut at 14:00 h), following a random order.

Forage was cut with a scythe, according to treatments it was wilted on the field, thereafter gathered and carried 300 m to the ensiling facility where it was cut into 3 cm (on average) particles using a Mapusa ® (Pudong, Shanghai, China) mincer. During mincing the forage was covered with a polyethylene film to avoid dehydration and contamination.

The inoculant used was BIOTAL PLUS II® (Lallemand Animal Nutrition, Milwaukee, Wisconsin) that contained viable cells of the LAB *Pediococcus pentosaceus* 12,455 (homofermentative) and *Propionibacterium freudenreichii* R2453 (heterofermentative) and specific enzymes, which were expected to enhance fiber hydrolysis during ensiling [5]. The inoculant was applied following instructions; hence, it was expected to supply 100,000 CFU/gram of forage.

For the compaction of the forage a structure was designed consisting of two vertical concrete cylinders of 0.25 m in diameter and 2.8 m in height, separated 1.15 m and joined at the top by a steel crossbar of 7.62 cm diameter and 1.25 m long. A hook with a pulley was placed on the crossbar, which was used to compact forage, operating with a rope to vertically move a cement piston with a diameter of 49 cm, height 25 cm and weight of 53 kg. The controlled displacement of the piston inside the containers was stabilized by placing them inside a metal ring 40 cm high that had hinges to open it when entering or removing each container.

The forage to be ensiled was placed within the containers in layers of 5 kg extended with pitchfork and the inoculant was applied with atomizer; two operators used the rope and the pulley to raise and drop the piston 60 times on each forage layer; when the section to be filled with the container was of less height than the piston, a metal sheet tube 50 cm high and 50 cm in diameter was used as a guide for the falling piston. Once each container was filled, the lid was placed and sealed with adhesive tape and secured with the strap. The containers were weighed on a Trutest® (Auckland, New Zealand) scale to verify that density was within target ranges (resulting in 608 ± 31 kg m^{-3}) and thereafter placed outdoors for 60 days, after which they were opened for sampling.

2.1. Measured variables

With a Davis Instruments Vantage Pro2® (Hayward, California) meteorological station, temperature, rainfall and humidity data were recorded at the time of forage harvesting, wilting and ensiling; these data were used to unravel the effect of weather variables on silage properties. There were no clear differences among phases with on average 0.14 mm rainfall in the morning hours and much less in the afternoon hours (0.04 mm); on the contrary, evapotranspiration was low in the morning hours (on average 0.21mm) and higher in the afternoon hours (on average 2.54 mm). The containers were opened 60 days after ensiling, a silage top layer of 30 cm was removed, the temperature was measured, and samples were taken for determinations of pH, DM content, aerobic deterioration and chemical composition.

2.1.1. Morphological characterization and quality indicators of the forage before ensiling

In the forage to be ensiled samples were taken to measure botanical composition, and DM content, temperature, pH, ASC (alcohol soluble carbohydrates), and buffering capacity.

Botanical and morphological composition of forage was estimated by means of hand separation. The DM content was estimated by drying at 65°C to constant weight, this variable was estimated in freshly cut forage or after wilting according to treatments, and additionally at the beginning, half and end of the process of filling the containers.

Temperature and pH measurements were made with a portable Orion 3-Star® meter (Thermo Fisher Scientific Inc., Chelmsford, Massachusetts). The temperature was measured at the beginning, middle and end of the container filling process. For the pH measurement the samples of approximately 50 g of fresh forage were taken in the field, they were frozen and subsequently the pH was measured in the laboratory. The determination of ASC was carried out in 40–50 mg of sample previously ground using the method of Dubois et al. [9]. The buffer capacity of the fresh forage was determined in 2.5 g fresh samples according to the method described by Jasaitis et al. [10].

2.1.2. Variables measured in the silage

The measurements made in the silage comprised the following variables: (i) temperature, pH and ASC content, such as in fresh forage, (ii) aerobic deterioration, (iii) crude protein content (CP) and neutral detergent fiber (NDF), (iv) rate of ruminal fermentation and *in vitro* disappearance of DM, and (v) NH3-N content.

At the opening of each silo, the temperature of the silage was measured at five points of the surface layer, and five points at a depth of 30 cm; 100 g samples were taken, which were refrigerated for pH measurement in the laboratory 4 h later. Likewise, samples were taken to which the ASC content was determined with the same procedure used in fresh forage samples.

The aerobic deterioration of the silage was estimated with measurements of temperature and CO2 production. Three 300 g silage samples were placed in 1 L transparent glass jars. Two of those jars were used to measure CO_2 production following Crossno et al. [11]; the third one remained 5d uncovered and was used to measure the temperature at 08:00, 14:00 and 20:00 h.

The CP content of the silage was estimated by the Kjeldahl method [12], while to estimate its NDF content, the Van Soest method described by 13 Sosa [13] was used.

The ruminal fermentation of silage samples was estimated using the gas production technique [14] with three replicates per sample. The gas pressure generated by the fermentation was measured with a manometer with a scale of 0–1 kg cm^{-2}, equipped with a three-phase key and hypodermic needle, the measurements were made with intervals of one-hour in the first 24 h and then every 4 h, after each reading the pressure was made equal to zero. The total gas production was estimated and once the incubation period was over, the residue was filtered and dried, which was considered as the residual DM. The in vitro disappearance of DM (DM*iv*D) was calculated as the difference between initial DM and residual DM.

For the estimation of the NH_3-N content, metaphosphoric acid was added to the samples and NH_3 was quantified by ultra violet light chromatography with a visible ultraviolet light spectrophotometer (UV/VIS Lambda 35, Perkin Elmer, Waltham, Massachusetts) at 630 nm [15].

Data were analyzed using a generalized linear model (GLM procedure 16 SAS 9.0®) which corresponded to the complete factorial arrangement and hence with effects of cutting schedules, LAB inoculant, wilting times, double interactions, phases and replicates within phases, the Tukey procedure ($\alpha = 0.05$) was used for comparisons of LS-means [16]. Additionally, a Principal Component Analysis was carried out with Microsoft Excel® using XLSTAT® (Addinsoft, New York, NY).

3. Results

The average pH of the fresh forage was 6.9 ± 0.3 and its buffering capacity was equivalent to 339 ± 50 meq 10^{-3} lactic acid, with no effect of the combinations of experimental phases, harvesting hours, wilting times and DM content of the forage at the time of ensiling.

3.1. Dry matter content of forage at cutting and post-wilting

The DM content of forage wilted for 2 h did not differ ($p < 0.05$) from that of the forage wilted for 1 h. The forage cut at 14:00. had a higher DM content than that cut at 08:00 h in Phase 1 ($p = 0.0029$) and in Phase 2 ($p = 0.0065$), but not in Phase 3 ($p = 0.3348$) (**Table 1**). No differences ($p > 0.05$) were found in DM content between wilting 1 or 2 h. The post-wilting DM content was not affected by wilting in phase one; but, in phases two and three the forage after 1 or 2 h of wilting, cut at 2:00 PM had a higher DM content than that cut at 08:00 h under the same wilting time ($p < 0.05$) (**Table 1**).

3.2. Contents of crude protein, neutral detergent fiber and alcohol-soluble carbohydrates in forage to be ensiled

Hour of cutting and wilting time did not affect ($p > 0.05$) CP content in the three experimental phases; contents were on average 18.5 ± 1.2, 16.7 ± 0.7 y 16.0 ± 0.6. In phase one the content of ASC was not affected ($p > 0.05$) by hour of cutting or wilting time, while in phases two and three it was higher ($p < 0.05$) in forage harvested at 14:00 (**Table 2**).

	Phase 1	Phase 2	Phase 3
Hour at cutting	Freshly cut forage DM(%)		
8:00	23 ± 1.1[b]	17 ± 0.7[b]	20 ± 1.1[a]
14:00	29 ± 1.1[a]	21 ± 0.7[a]	22 ± 1.1[a]
	Post-wilting DM (%)		
8:00	27 ± 2[a]	20 ± 2[b]	19 ± 1[b]
14:00	27 ± 2[a]	24 ± 2[a]	26 ± 1[a]

[ab] Means with different literal within columns (of freshly cut or after wilting) are different ($p \leq 0.05$).

Table 1. Dry matter content (DM) of alfalfa and orchard grass herbage under two harvest hours, freshly cut and after one or 2 h of wilting during three experimental phases (LS-means).

		ASC (%)			NDF (%)		
Hour at cutting	Wilting time (h)	Phase 1	Phase 2	Phase 3	Phase 1	Phase 2	Phase 3
8:00	0	3.8[a]	3.4[b]	4.3[b]	57[d]	75[a]	69[b]
8:00	1	3.5[a]	3.1[b]	4.0[b]	63[b]	75[a]	73[a]
8:00	2	3.3[a]	3.3[b]	4.3[b]	68[a]	75[a]	66[c]
14:00	0	4.3[a]	3.9[a]	4.7[a]	60[c]	73[b]	67[bc]
14:00	1	5.1[a]	4.5[a]	4.5[a]	61[bc]	71[c]	67[bc]
14:00	2	3.8[a]	4.2[a]	4.9[a]	57[d]	71[c]	63[d]
SE of means		0.8	0.2	0.2	2.6	1.9	2.1

[abcd] Means that do not share any literal within columns are different ($p \leq 0.05$).

Table 2. LS-means and standard error (SE) of alcohol soluble carbohydrates (ASC) and NDF alfalfa and orchard grass forage before ensiling.

3.3. Temperature, pH and ammoniacal nitrogen of silages

The average temperature at the opening of the silos was 21 ± 0.6°C in phase one, 20 ± 0.7°C in phase two and 18 ± 1.1°C in phase three, not affected by treatments. In phase one, the silage pH of forage harvested at 14:00 and with inoculant was lower ($p < 0.05$) than that of forage harvested at the same time but without inoculant and those of forage harvested at 08:00 (**Table 3**). The NH3-N content was not affected ($p > 0.05$) by treatments in phases one and two. In phase three the NH3-N content of forage harvested at 14:00 and with inoculant was lower ($p < 0.05$) than that of forage harvested at the same time but without inoculant and those of forage harvested at 08:00 (**Table 3**).

3.4. Alcohol-soluble carbohydrates in silages

The ASC in freshly cut forage was reduced by ensiling and hence was lower in the correspondent silages, in such a way that there was a linear relationship ($p < 0.01$) between ASC in

		pH			N-NH_3 (% of total N)		
Hour at cutting	Inoculant	Phase 1	Phase 2	Phase 3	Phase 1	Phase 2	Phase 3
8:00	0	4.7^b	4.9^a	4.8^a	11^a	16^a	15^b
8:00	1	4.8^a	4.9^a	4.9^a	11^a	14^a	18^a
14:00	0	4.8^a	4.6^b	4.6^b	11^a	14^a	13^c
14:00	1	4.4^c	4.6^b	4.7^b	9^a	14^a	12^c

^{abc} Means that do not share any literal within columns are different ($p \leq 0.05$).

Table 3. Means pH and ammoniacal nitrogen of silages alfalfa and orchard grass cut at two different hours, with or without addition of bacterial inoculant, in three distinct experimental phases.

freshly cut forage and the reduction in content due to ensiling (**Figure 1**). There were significant differences between the treatments in phase two because of the wilting time factor, where the treatments without wilting maintained an ASC of 1.2%, with one-hour of wilting was 0.8% and with 2 h of wilting 1.1%.

3.5. Nutritional composition of silages

The CP content of silages was not affected ($p > 0.05$) by the treatments, it was on average 18% in phase one, 15% in phase two and 16% in phase three. The differences in CP of silage with respect to freshly cut forage were small, in phase one the protein of the silage was 0.5% lower, in phase two the decrease was 1.7% while in phase three contents were similar.

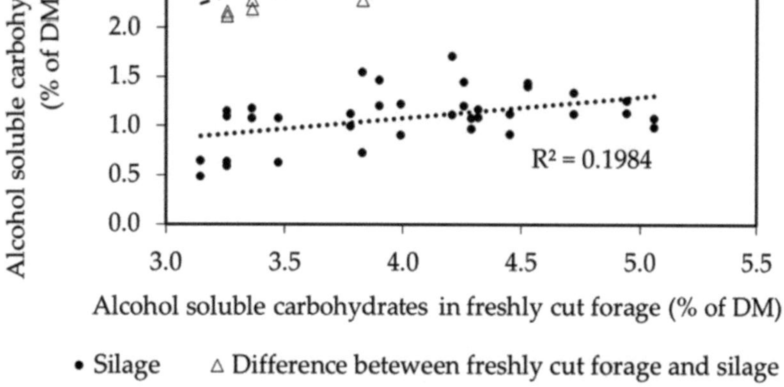

Figure 1. Relationship between the content of alcohol-soluble carbohydrates in the original forage and its corresponding silage.

In all three phases the NDF content decreased 10% in silage with respect to that of freshly cut forage. In phases one and three the NDF content in the silage was not affected by the treatments ($p > 0.05$). In phase two the hour of cutting x wilting time interaction led to differences ($p < 0.05$); while silage from forage harvested at 8:00 with one-hour wilting and at 14:00 with 2 h wilting had the highest contents, those from forage harvested at 8:00 with 2 h wilting and at 14:00 with one-hour wilting had the lowest contents (**Table 4**).

3.6. Fermentation and ruminal in vitro disappearance of DM

The volume of gas produced by in vitro silage fermentation was not affected by the treatments in phase one ($p > 0.05$); however, in phase two the volume of gas produced by the fermentation of silage from forage harvested at 08:00 h was lower ($p < 0.05$) than the volume reached with forage harvested at 14:00 (**Table 5**). In phase three, the gas volumes were affected ($p < 0.05$) by the interaction of wilting time x inoculant; treatments without wilting and with 2 h of wilting with inoculant application reached higher gas volumes than treatments without inoculant. In the case of treatments with one-hour of wilting, the upper value of gas volume was achieved in treatments without inoculant. The *in vitro* disappearance of DM did not show significant differences between treatments of phases one and two; on the other hand, in phase three the percentage of disappearance was lower ($p < 0.05$) in silages from forage harvested at 08:00 h (**Table 5**).

3.7. Aerobic stability of silages

The measurement of the rate of change of temperature did not produce clear results. In phase one the interactions hour at cutting x inoculant ($P < 0.05$) on the rate of change of temperature resulted in silage from forage harvested at 8:00 with one-hour wilting and at 14:00 with 2 h wilting the highest rates were found, while in silages from forage harvested at 8:00 with 2 h wilting and at 14:00 with one-hour wilting the lowest rates were detected. In the same first phase the results of the wilting time x inoculant interaction were of such a nature that no

Hour at cutting	Wilting (h)	NDF (% of DM)		
		Phase 1	Phase 2	Phase 3
8:00	0	47^{a}	59^{b}	63^{a}
8:00	1	48^{a}	61^{a}	64^{a}
8:00	2	45^{a}	54^{e}	60^{a}
14:00	0	48^{a}	57^{c}	62^{a}
14:00	1	50^{a}	56^{d}	65^{a}
14:00	2	51^{a}	61^{a}	59^{a}

^{abcde} Means that do not share any literal within columns are different ($p \leq 0.05$).

Table 4. Means of NDF content of silage from alfalfa and orchard grass forage harvested at two distinct times of the day and subjected to different times of wilting during three experimental phases.

Hour at cutting	Phase 1	Phase 2	Phase 3
	Gas volume (ml g^{-1} DM)		
08:00	218[a]	193[b]	227[a]
14:00	223[a]	209[a]	238[a]
	Ruminal *in vitro* disappearance of DM (%)		
08:00	60[a]	49[a]	50[b]
14:00	58[a]	50[a]	54[a]

[ab] Means that do not share any literal within columns are different ($p \leq 0.05$).

Table 5. Means of volume of gas produced by fermentation and ruminal in vitro disappearance of DM of silage from alfalfa and orchard forage harvested at contrasting times of the day in three experimental phases.

rational interpretation was feasible. In phases two and three no effect ($P > 0.05$) of treatments on this variable was detected. Differences among means of phases one, two and three (0.17, −0.27, 0.03°C d^{-1}) were also odd.

Concerning the variable CO_2 production rate, no effects of treatments were detected in any of the three phases($P > 0.05$); the rates of CO_2 production were 9, 9.4 and 19.8 milli moles of CO_2 g^{-1} DM d^{-1} in phases one, two and three, respectively. In the rate of change of pH, the interaction Hour at cutting x Inoculant (**Table 6**) in the three phases implied that inoculation was not effective to control pH rise in silage from forage cut at 08:00 but it was on silage from forage cut at 14:00 ($p < 0.05$). The interaction Inoculant x wilting time (**Table 6**) in phase one

		Rate of change of pH units d^{-1}		
Hour at cutting	Inoculant	Phase 1	Phase 2	Phase 3
08:00	0	0.18[d]	0.86[a]	0.18[a]
08:00	1	0.63[a]	0.95[a]	0.42[a]
14:00	0	0.37[b]	0.58[b]	−0.002[b]
14:00	1	0.25[c]	0.32[b]	0.15[a]
Inoculant	Wilting time	Phase 1	Phase 2	Phase 3
0	0	0.33[d]	0.91[a]	0.12[a]
0	1	0.10[e]	0.77[a]	0.05[a]
0	2	0.40[c]	0.49[a]	0.09[a]
1	0	0.04[f]	0.72[a]	0.30[a]
1	1	0.84[a]	0.65[a]	0.57[a]
1	2	0.45[b]	0.52[a]	0.00[a]

[abcde] Means that do not share any literal within columns of each interaction are different ($p \leq 0.05$).

Table 6. Means of rate of change of pH during aerobic deterioration of alfalfa and orchard grass silages.

without inoculant, one-hour wilting precluded the rise of pH, but two-hour wilting did not ($p < 0.05$), while with the addition of inoculant, wilting led to rises in pH ($p < 0.05$), in phases two and three there was no effect of this interaction ($p > 0.05$).

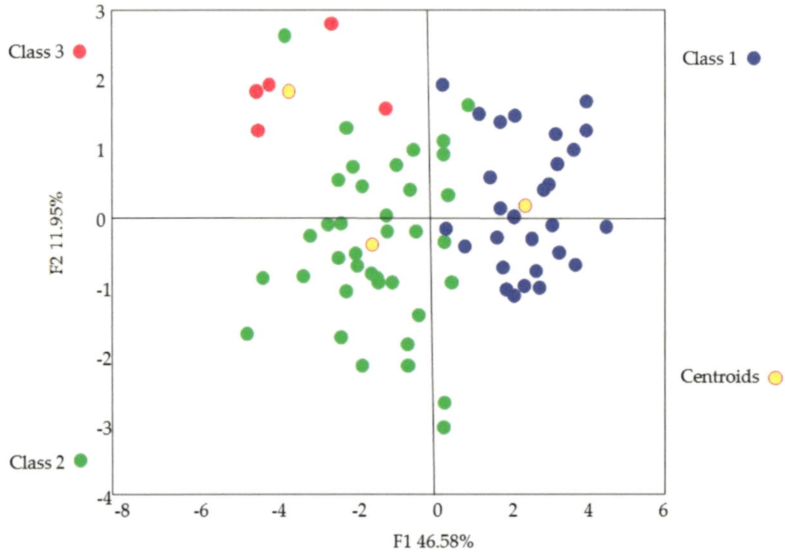

Figure 2. Classes of alfalfa and orchard grass silages harvested at 08:00 or 14:00, submitted to 0, 1 or 2 h of wilting and with or without addition of bacterial inoculant.

Class	1	2	3
Harvest at 14:00 proportion (%) within class	89%	26%	17%
Harvest at 08:00 proportion (%) within class	11%	74%	83%
Wilting	75%	58%	83%
Inoculant	46%	50%	67%
Events rainfall	11%	16%	100%
Amount rainfall (mm/6 h)	0.3	0.2	0.7
Evapotranspiration (mm/6 h)	2.5	0.7	0.3
DM % Freshly cut forage	25%	20%	19%
DM % Silage	29%	20%	18%
pH	4.56	4.80	5.18
CO_2 Production rate mmol CO_2 g^{-1} DM d^{-1}	5.8	17.2	16.2
Temperature °C	20.4	19.1	19.9

Table 7. Attributes of three classes of alfalfa and orchard grass silages harvested at 08:00 or 14:00, wilted 0, 1 or 2 h and with or without addition of bacterial inoculant.

3.8. Results of principal component analysis

Three classes of silages were identified; Class 1, Class 2 and Class 3 comprising 39%, 53% and 8% of silages (**Figure 2**). Characteristics of those classes are described in **Table 7**.

4. Discussion

4.1. Properties of forage to be ensiled

The pH of forage to be ensiled was not affected by treatments and ranged between 6.5 y 7.0 in the scope of results quoted by Coblentz and Muck [4], but somewhat higher than the range between 6.1 and 6.2 reported by Santos and Kung [17]. The results attained in terms of buffering capacity with no effect of treatments, differ from those reported by Zheng et al. [18] who found that wilting resulted in an on average reduction in 5% of the buffering capacity of alfalfa forage; but, the effect of wilting on the buffering capacity is highly dependent on the weather conditions during wilting [4].

As expected from results reported by Tremblay et al. [3] DM content of forage cut in the afternoon was higher than that of forage cut in the morning, which was coupled with higher humidity and lower ambient temperature, solar radiation and wind speed in the morning hours, leading to lower evapotranspiration which concurs with Owens et al. [19]. During morning hours low evapotranspiration is the probable cause of the lack of effect of wilting on DM content of forage to be ensiled; conversely, in forage harvested at 14:00 in phases two and three there was a linear increase in DM content as the wilting time increased.

The content of ASC of fresh forage in phase one did not show significant differences, while in phases two and three forage treatments harvested at 14:00 had higher concentrations of ASC, which was to be expected [3]. On the other hand, the wilting time did not affect the concentration of ASC; on the contrary, Zheng et al. [18] found reductions of 8 and 17% when wilting alfalfa forage for 2 and 4 h.

The NDF content of forage harvested at 14:00 was lower than that of forage harvested at 08:00, probably due to the accumulation of photosynthetic products in the cell content [19]. This represents advantages in terms of the nutritional composition and fermentative characteristics of the original forage [3].

4.2. Characteristics of silages

Silage temperatures (in the range between 18 and 21°C) were adequate, Borreani and Tabacco [20] report that in well preserved silage the temperature should be close to the ambient temperature.

Silage pH is one of the main factors that influences the degree of proteolysis [21] in silages; the results attained (**Table 3**) were not conclusive since treatments effects differed between phases and short-term weather variables are a feasible explanation for these differences. The pH values fluctuated between 4.6 and 4.9; analogously, in alfalfa silage with *L. buchneri* inoculant

[22] detected a pH of 5.0. In phase two, the pH values of silage from forage harvested at 14:00 were lower, which would coincide with higher levels of MS and ASC contents, conditions that optimize LAB activity and explain the advantage of harvesting forage in the afternoon hours [19]. According to Tyrolová and Výborná [23] for forages with less than 20% DM, it is necessary to acidify the forage up to a minimum pH of 4.2 and for forages with 30% DM a pH of 4.45 is acceptable, which implies that average pH in silages of this experiment were somewhat high. Concurrently, Kung [24] states that pH values higher than 4.6 to 4.8 in legume silages may be due to ensiling at DM contents lower than 30%.

The NH3-N content is an important indicator, since it shows the amount of protein that has been degraded to ammonia. High NH_3-N contents such as found in this experiment (except for silages from forage harvested at 14:00 and with inoculant added) result from extremely high breakdown of protein, which is frequent in silages with DM contents lower than 30% [24].

The concentration of ASC in the original forage was on average 4.1% and was reduced in all silages in increasing proportion as the initial content was higher ($P < 0.01$), indicating that they were efficiently used by LAB [21]. If there had been a higher concentration of ASC in the original forage, the silage pH would have been even lower; therefore, according to the results of the review by Yitbarek and Tamir [25], the addition of a source of highly fermentable carbohydrates such as molasses would be a suitable alternative.

In the three phases the content of NDF in silage decreased as the wilting time increased, an effect that coincides with the results of Hashemzadeh-Cigari et al. [26] who found that wilting alfalfa before ensiling decreased the NDF content in silage.

The in vitro disappearance of DM presented average values of 59, 50 and 52% in phases one, two and three respectively and are lower than the values reported by Rizk et al. [27] who reported an average of 65%, these differences can possibly be attributed to the fact that in the present study the ensiled forage had low DM content.

4.2.1. Aerobic stability of silages

Forage silages harvested at 08:00 h and inoculated were less stable than those ensiled without inoculants; on the other hand, in forage harvested at 14:00, inoculated silages were more stable. Similarly inoculants were more effective in improving different silage attributes when applied to wilted forage than to fresh forage [26]. Improvement of the aerobic stability characteristics of silages, as in the present study, might be expected with the use of heterofermentative LAB inoculants or mixed heterofermentative and homofermentative LAB [7].

4.3. Results of principal component analysis

In Class 1 the best quality silages were found, with lower pH, lower aerobic deterioration and higher DM content, the silages of that class were, in a very high proportion, harvested at 14:00 with very low proportion of rain events, relatively high evapotranspiration, mostly subjected to wilting (75% versus 66% expected), with highest DM% of original forage, and as in the other two classes no clear effect of inoculant addition. The highest proportion of silages was identified in Class 2, of lower quality than Class 1. The main differences between these two classes were harvest time (Class 2 mostly harvested at 08:00) and, during the previous 6 h to sealing of silos, 72% lower evapotranspiration and 47% higher proportion of rain events. Class 3 comprised a low proportion of spoiled silage which faced heavy rainfall in the 6 h prior to the closure of silos.

From the above, it follows that the harvest in the afternoon after consulting weather forecasts trying to ensure that there will not be rain events and that there will be conditions for high evapotranspiration are essential factors for the success of ensiling; these results concur with those of Coblentz and Muck [4]. If these conditions are met, wilting and inoculation can contribute to improve the silage quality.

5. Conclusions

Harvesting alfalfa and orchard grass forage at 14:00 was advantageous since it led to silage with higher contents of dry matter and alcohol soluble carbohydrates and lower content of neutral detergent fiber.

Absence or rain and high evapotranspiration favored the achievement of positive effects of harvesting in afternoon hours. The effect of wilting and inoculation were bound to rainfall and evapotranspiration conditions during ensiling; under good weather conditions of wilting and inoculation contributed to improve the silage quality in terms of high dry matter content, low pH and better aerobic stability.

Acknowledgements

The authors are grateful to CONACYT for the M Sc Studies grant of the second author.

Conflict of interest

Authors declare they have no conflict of interest.

Author details

Ricardo D. Améndola-Massiotti[1]*, Renato González-Ortiz[1], Luis A. Miranda-Romero[1], Juan A. Burgueño-Ferreira[2] and Pedro Topete-Pelayo[3]

*Address all correspondence to: r_amendola@yahoo.com

1 Animal Science Graduate Program, Autonomous University Chapingo, Texcoco, Mexico

2 CIMMYT. International Maize and Wheat Improvement Center, El Batán, Mexico

3 Animal Science Department, Autonomous University Chapingo, Texcoco, Mexico

References

[1] Améndola R, Castillo E, Martínez P. FAO Country Pasture/Forage Resource Profiles. Mexico [Internet]. 2005. Available from: https://www.researchgate.net/publication/325859440_FAO_Country_PastureForage_Resource_Profiles_Mexico [Accessed: 20-06-2018]

[2] Wen AY, Yuan XJ, Wang J, Desta ST, Shao T. Effects of four short-chain fatty acids or salts on dynamics of fermentation and microbial characteristics of alfalfa silage. Animal Feed Science and Technology. 2017;**223**:141-148. DOI: 10.1016/j.anifeedsci.2016.11.017

[3] Tremblay GF, Morin C, Bélanger G, Bertrand A, Castonguay Y, Berthiaume R, et al. Silage fermentation of PM- and AM-cut alfalfa wilted in wide and narrow swaths. Crop Science. 2014;**54**:439-452. DOI: 10.2135/cropsci2013.07.0443

[4] Coblentz WK, Muck RE. Effects of natural and simulated rainfall on indicators of ensilability and nutritive value for wilting alfalfa forages sampled before preservation as silage. Journal of Dairy Science. 2012;**95**:6635-6653. DOI: 10.3168/jds.2012-5672

[5] Addah W, Baah J, McAllister TA. Effects of an exogenous enzyme-containing inoculant on fermentation characteristics of barley silage and on growth performance of feedlot steers. Canadian Journal of Animal Science. 2016;**96**:1-10. DOI: 10.1139/cjas-2015-0079

[6] Twarużek M, Dorszewski P, Grabowicz M, Szterk P, Grajewski J, Kaszkowiak J. Effect of additives on the fermentation profile, microbial colonization and oxygen stress of alfalfa silages. Journal of Elementology. 2016;**21**:1161-1172. DOI: 10.5601/jelem.2015.20.4.1066

[7] Kung L Jr. Aerobic stability of silage. In: Proceedings of the 2010 California Alfalfa & Forage Symposium and Corn/Cereal Silage Conference; 1-2 December 2010; Visalia, CA. [Internet]. 2010. Available from: http://alfalfa.ucdavis.edu/+symposium/proceedings/2010/10-89.pdf [Accessed: 15-06-2018]

[8] Li M, Shan G, Zhou H, Fan Y, Ma D, Wang Z, et al. CO_2 production, dissolution and pressure dynamics during silage production: Multi-sensor-based insight into parameter interactions. Scientific Reports. 2017;**7**:14721. DOI: 10.1038/s41598-017-14187-1

[9] Dubois MK, Gilles A, Hamilton JK, Rebers PA, Smith F. Colorimetric method for determination of sugars and related substances. Analytical Chemistry. 1956;**28**:350-356. DOI: 10.1021/ac60111a017

[10] Jasaitis DK, Wohlt JE, Evans JL. Influence of feed ion content on buffering capacity of ruminant feedstuffs in vitro. Journal of Dairy Science;**70**:1391-1403

[11] Crossno SK, Kalbus LH, Kalbus GE. Determinations of carbon dioxide by titration: New experiments for general, physical, and quantitative analysis courses. The Chemical Educator. 1996, 1987;**73**:175-176. DOI: 10.1021/ed073p175

[12] Horwitz W, editor. Official Methods of Analysis of AOAC International. 17th ed. Gaithersburg Maryland: AOAC International; 2000. p. 2200

[13] Van Soest PJ, Robertson JB, Lewis BA. Methods for dietary fiber, neutral detergent fiber, and nonstarch polysaccharides in relation to animal nutrition. Journal of Dairy Science. 1991;**74**:3583-3597. DOI: 10.3168/jds.S0022-0302(91)78551-2

[14] Menke KH, Steingass H. Estimation of the energetic feed value obtained from chemical analysis and in vitro gas production using rumen fluid. Animal Research and Development. 1988;**28**:7-55

[15] McCullough H. The determination of ammonia in whole blood by direct colorimetric method. Clinica Chimica Acta. 1967;**17**:297-304. DOI: 10.1016/0009-8981(67)90133-7

[16] SAS Institute. SAS User's Guide: Statistics. Ver. 9.2. Cary, N.C.: SAS Institute; 2004. p. 5180

[17] Santos MC, Kung L Jr. Short communication: The effects of dry matter and length of storage on the composition and nutritive value of alfalfa silage. Journal of Dairy Science. 2016;**99**:5466-5469. DOI: 10.3168/jds.2016-10866

[18] Zheng M, Niu D, Zuo S, Mao P, Meng L, Xu C. The effect of cultivar, wilting and storage period on fermentation and the clostridial community of alfalfa silage. Italian Journal of Animal Science. 2018;**17**:336-346. DOI: 10.1080/1828051X.2017.1364984

[19] Owens VN, Albrecht KA, Muck RE. Protein degradation and fermentation characteristics of unwilted red clover and alfalfa silage harvested at various times during the day. Grass and Forage Science. 2002;**57**:329-341. DOI: 10.1046/j.1365-2494.2002.00333.x

[20] Borreani G, Tabacco E. The relationship of silage temperature with the microbiological status of the face of corn silage bunkers. Journal of Dairy Science. 2010;**93**:2620-2629. DOI: 10.3168/jds.2009-2919

[21] Filya I, Sucu E, Karabulut A. The effect of *Lactobacillus buchneri* on the fermentation, aerobic stability and ruminal degradability of maize silage. Journal of Applied Microbiology. 2006;**101**:1216-1223. DOI: 10.1111/j.1365-2672.2006.03038.x

[22] Jatkauskas J, Vrotniakiene V, Ohlsson C, Lund B. The effects of three silage inoculants on aerobic stability in grass, clover-grass, lucerne and maize silages. Agricultural and Food Science. 2013;**22**:137-144. Retrieved from https://journal.fi/afs/article/view/6698

[23] Tyrolová Y, Výborná A. The effects of wilting and biological and chemical additives on the fermentation process in field pea silage. Czech Journal of Animal Science. 2011; **10**:427-432. DOI: 10.17221/3235-CJAS

[24] Kung L Jr. Understanding the biology of silage preservation to maximize quality and protect the environment. In: Proceedings of the 2010 California Alfalfa & Forage Symposium and Corn/Cereal Silage Conference; 1-2 December 2010; Visalia, CA. [Internet]. 2010. Available from: http://alfalfa.ucdavis.edu/+symposium/proceedings/2010/10-89.pdf [Accessed: 15-06-2018]

[25] Yitbarek MB, Tamir B. Silage additives: Review. Open Journal of Applied Sciences. 2014; **4**:258-274. DOI: 10.4236/ojapps.2014.45026

[26] Hashemzadeh-Cigari F, Khorvash M, Ghorbani GR, Taghizadeh A. The effect of wilting, molasses and inoculants on the fermentation quality and nutritive value of lucerne silage. The South African Journal of Animal Science. 2011;**41**:388-378. DOI: 10.4314/sajas.v41i4.8

[27] Rizk C, Mustafa AF, Phillip LE. Effects of inoculation of high dry matter alfalfa silage on ensiling characteristics, ruminal nutrient degradability and dairy cow performance. Journal of the Science of Food and Agriculture. 2005;**85**:743-750. DOI: 10.1002/jsfa.2034

Chapter 8

Tropical Forage Legumes in India: Status and Scope for Sustaining Livestock Production

Tejveer Singh, Srinivasan Ramakrishnan,
Sanat Kumar Mahanta, Vikas C. Tyagi and
Ajoy Kumar Roy

Additional information is available at the end of the chapter

http://dx.doi.org/10.5772/intechopen.81186

Abstract

Livestock contributes enormously in food and nutritional security apart from livelihood security to rural population all over the world. India has the largest number of livestock, representing over 17% of world population. Availability of forage legumes is essential for better animal health, production and increasing the nutritive value of forage-based rations, besides providing a source of biological nitrogen fixation for enriching soil, reducing land degradation and mitigating climate change. However, supply of quality green fodder in India is extremely precarious, and the gap is huge against demand. The major fodder legume crops cultivated in India are *Medicago sativa, Trifolium alexandrinum, Vigna unguiculata, Vigna umbellate* and range legumes are *Stylosanthes* spp., *Desmanthus virgatus*, and *Clitoria ternatea*. Indian subcontinent represents wide spectrum of eco-climates and reported diversity of 21 forage legumes genera *viz., Desmodium, Lablab, Stylosanthes, Vigna, Macroptelium, Centrosema* and browse plants *Leucaena, Sesbania, Albizia, Bauhinia, Cassia, Grewia*, etc. Diversity of forage legumes were collected (>3200 accessions), evaluated and sources for different biotic and abiotic stress tolerance were identified, apart from >50 cultivars developed. Considering these aspects, tropical legumes for livestock production, soil health and ecosystem services, diversity, evaluation and breeding for improved varieties are discussed in this chapter.

Keywords: crop wild relatives, gene introgression, germplasm, range legumes, livestock production, N-fixation

IntechOpen

1. Introduction

Cultivated forage legumes and range legumes are contributing in sustainable agriculture production apart from nutritional security to the livestock population of India. Cultivated forage legumes and range legumes are also crucial for the nutritional security for mankind as they are integral component for increased availability of animal protein and product which has higher biological value than the plant proteins. The major fodder legumes crops cultivated in India are *Medicago sativa, Trifolium alexandrinum, Vigna unguiculata, Mucuna pruriens, Vigna umbellate* and range legumes are *Stylosanthes* spp., *Desmanthus virgatus, Clitoria ternatea* and others. Among these, *Medicago sativa, Trifolium alexandrinum* and *Vigna unguiculata* are more popular among cultivated legumes and *Stylosanthes* in range legumes because of easy availability of seeds of improved varieties and well developed technology to increase the forage yield and quality. To understand the current status and scope of tropical forage legumes of India for sustaining income through livestock sector, their importance in livestock production, soil health and ecosystem services and diversity among germplasms, evaluation and breeding for improved varieties are discussed in this chapter.

2. Forage legumes in livestock production

India has the largest livestock population in the world with more than 512 million heads. It supports 56.7% of the world's buffaloes, 12.5% of the world's cattle and 20.4% of the world's small ruminants (sheep and goats) [1]. Besides, the country hosts 17% of the world human population [2]. India is also the leading milk producing country in the world but milk productivity per animal basis is very low. Deficiency in quality of fodder is one of the major reasons for the low animal productivity. Although India is very rich in varied flora and fauna but there is deficiency of quality green fodder to the tune of around 35%. The animals need proper feeding to meet their nutrient requirement to express their full genetic production potential.

In fact, the sustenance of Indian rural agricultural economy depends on crop and animal farming, the two key components of a mixed farming system. Although the contribution of agricultural sector in the Indian economy is steadily declining (from 36.4% in 1982–1983 to 14.1% in 2012–2013), it still contributes employment to over 50% of the work force [3]. The contribution of livestock sector to agriculture GDP has increased to more than 28% and is likely to increase further. In the recent past, the lifestyle of people has been changed with a marked shift in food habits towards milk, milk products and meat leading to increase in demand of livestock products. Economic scenario in animal husbandry is also changing with emergence of peri-urban livestock farming and fodder markets. This indicates the huge pressure on available land, most of which, is used for arable farming and food production.

Forages form the main stay of our animal farming to reduce the competition between human beings and animals due to increasing demand for land and other inputs. Sole feeding of green forages to dairy animals is much cheaper than feeding concentrates with crop residues and has the potential of higher level of milk production. Nearly 65% of the total expenditure of milk production in cows is attributed to the feeding of animals when both concentrates and green fodders are fed as mixed ration. When the milk production is primarily depend upon concentrate based feeding, the cost of feeding towards milk production reaches to 80%, however, in case of forage (legumes) based feeding, it is reduced to only 40% of the total expenditure [4]. Hence, any attempt towards enhancing availability of quality green fodder, and economizing the feed cost would result in better remuneration to livestock farmers/producers.

From an animal perspective, one of the largest benefits provided by legume forages is that they provide a better level of nutrition than cereal forages/grasses at a similar stage of growth, leading to greater forage intake by livestock and increased animal performance. The symbiosis between legumes and Rhizobia provides the plant with an ample supply of N and it is one of the reasons why crude protein (CP) concentrations of legumes are higher than cereals/grasses. In addition to higher concentrations of CP, forage legumes also provide a higher quality protein which may be of equal or greater importance in case of non-ruminant livestock species like equines. Legumes also contain more concentrations of digestible energy than grass/cereal forages due to the structure and development of the legume cell wall. Indeed, the cell wall of legume plants contains fewer hemicelluloses and more pectin compared to that of cereals, thus increasing their digestibility by livestock. However as the cell matures, a secondary cell wall consisting of cellulose and lignin is deposited on the interior of the primary cell wall and reduces the overall availability of the structural carbohydrates in the digestive system. In cereal forages, this phenomenon occurs in all tissues types (i.e. leaves, stems, etc.) while being primarily restricted to the vascular tissues of legume stems. The lignin of non-legumes is also more esterified to hemicelluloses and is more recalcitrant in composition (e.g. higher proportion of syringyl subunits) indicating a more suppressed degradability than in legume species.

3. Forage legumes in soil health and ecosystem services

Forage legumes is essential for providing a source of biological nitrogen fixation (BNF) for enriching soil fertility (15–40 kg fixed N/ha), reduction in land degradation, disease breaks and for mitigating climate change. Estimating biological N_2 fixation of the forage and fodder legumes precisely is challenging because statistics on the areas and productivity of these legumes are highly difficult to obtain. Therefore, N_2 fixation values of forage and fodder legumes will be less reliable and also estimates of %Ndfa (nitrogen derived from atmosphere) of fodder legumes in those lands. There are very few reports available on forage legumes—BNF in India. But, all works mainly focused on application of *Rhizobium* inoculants to fodder legumes and testing their potential for enhancing fodder production (fresh and dry weight, crude protein content, forage quality aspects, nodulation properties, etc.). Appreciable amount of atmospheric N (~60–100%) is fixed by forage legumes annually, fixing up to 380 kg N ha^{-1} [5]. Quantity of forage residues available for soil incorporation range from 80 to 143 kg N ha^{-1} and rice cultivated following forage legumes yields the same as rice with 24–50 kg fertilizer N ha^{-1} [6]. About 100–120 Mha of land is under fodder and forage legumes and green manure crops, with assumed average N_2 fixation rates of 200 kg N/ha/year for alfalfa, 150 kg N/ha/year for clovers (*Trifolium* spp.), 100 kg N/ha/year for other forages and 50 kg N/ha/year for legume-grass pastures [7]. From this assumption, total nitrogen fixation by forage and fodder legumes was calculated at 12 Tg annually (average of about 110 kg N/ha/year). But fixation by legume-grass mixtures is much more variable, ranging from a just a few kilograms to more than 250 kg N ha^{-1}.

In India, area under fodder legumes and grasses is about 8 Mha (*Sorghum bicolor*—2.6 Mha, *Trifolium*—1.9 Mha, *Medicago*—1 Mha, other legume forages—1.9 Mha). Mean N uptake by *Trifolium alexandrinum* (240–264 kg/ha), *Medicago sativa* (216–264 kg/ha), *Vigna unguiculata* (161–181 kg/ha), *Sorghum bicolor* (128–160 kg/ha), BN hybrid (*Pennisetum glaucum* × *Pennisetum purpureum*) and *Megathyrsus maximus* (288–360 kg/ha), *Avena sativa* (120–144 kg/ha). Percent nitrogen derived from atmosphere (%Ndfa) is about 0.7 for legumes and 0.1 for cereals/grasses. Annual contribution of BNF by forage and fodder crops in India is about 0.61 Tg/year which is nearly 5% of world BNF of forage and fodder [8]. However, majority of values available for legume N_2 fixation were based on shoots and above ground parts only. They did not include the fixed N present in roots, nodules and rhizodeposition in general. Published values for below-ground N as a percentage of the total plant N are 22–68% for the pulse and oilseed legumes, *Glycine max*, *Vicia faba*, *Cicer arietinum*, *Vigna radiata*, *Lupinus albus*, *Pisum sativum* and *Cajanus cajan* and 34–68% for the pasture/fodder legumes, subterranean clover, white clover and alfalfa [9–11].

In addition to BNF, many forage legumes have soil-covering growth habit similar to most grasses and deep root system which can contribute to the mitigation of many soil problems, *viz.*, soil conservation by legume cover crops such as *Stylosanthes*, *Crotalaria*, *Sesbania*, *Arachis* and *Desmodium* to prevent erosion; contour-hedges with leguminous trees such as *Leucaena*; rehabilitation of degraded soils by legumes such as *Stylosanthes* spp., which are deep-rooted and adapted to infertile soils, cycle minerals from deeper soil layers resulting in soil improvement and enhanced concentration of soil organic matter through litter production [12]; the potential of legumes like *Stylosanthes hamata* can be exploited to ameliorate compacted soil [13]. When used as cover crop forage legumes can also control weed growth, which can be exploited as an attractive alternative to the use of herbicides. They supplement part of N fertilizer application, thus reduce nitrate leaching and eutrophication of water bodies as a consequence of surface runoff as a result of N fertilization in tropical pasture production process. Tropical forage legumes have considerable potential to increase productivity of forage-based livestock systems, while providing benefits to the environment [14]. The environmental benefits, referred as 'ecosystem services', comprise positive effects on: soil conservation and soil chemical, physical and biological properties; mitigation of global warming and of groundwater contamination; saving of fossil energy; and rehabilitation of degraded lands [14]. These features make tropical forage legumes particularly valuable at all levels of the system because of their interaction with plants, soil, animals and the atmosphere.

4. Genetic resources of tropical forage legumes

Plant genetic resources (PGR) are the basic platform for screening, improving and developing fine cultivars, and the important materials for biodiversity studies including classification, evolution and origin. Therefore, maintenance of enormous genetic diversity is mandatory for broadening the genetic base of the present and future forage improvement programmes to achieve the national goals. Extensive collection, proper evaluation, in depth study of genetic attributes and cataloging of germplasm is prerequisite for its efficient utilization. According to an estimate there are about 650 genera, 18,000 species of legumes (Leguminosae) in the world. Out of these, only about 30 legumes are used to an appreciable extent for forage production [15]. Information regarding the centre of origin of different forage crops is furnished in **Table 1**.

World-wide, 1500 gene banks are registered in the WIEWS (World Information and Early Warning System on PGR) database [16] and conserve a total of 7.1 million accessions belonging to 53,109 species, including major crops, minor or neglected crop species, as well as trees and wild plants. Out of total germplasms stored, 651,024 accessions belonging to forage

Genus	Species	Centre of origin	Distribution
Atylosia	*scarabaeoides*	India	
Centrosema	*pubescens*	South America	South east Asia, Indonesia and Africa
Clitoria	*ternatea*	Tropical America	Tropical and subtropical parts of the world
Desmanthus	*virgatus*	Argentina	Florida, throughout the India
Desmodium	*intortum*	Central and South America	Throughout the tropical areas of Africa, Australia and new world
Macroptilium	*atropurpureum*	Central and South America	Australia, South east Asia, Pacific Islands
Macroptilium	*lathyroides*	India	Tropical and subtropical world
Macrotyloma	spp.	Africa and Asia	Sri Lanka
Macrotyloma	*uniflorum*	India	Africa
Stylosanthes	*guianensis*	Brazil	West Indies, Africa and Pacific Islands
Stylosanthes	*hamata*	Islands of West Indies	Coastal regions of north and south America
Stylosanthes	*humilis*	North east Brazil and Venezuela	Tropical parts of world
Stylosanthes	*scabra*	Tropical America	Kenya, Brazil and Queensland
Stylosanthes	*seabrana*	Brazil	
Lablab	*purpureus*	Asia or Africa	India, subtropical areas of Africa, south Asia
Cyamopsis	*tetragonoloba*	Africa	India (secondary centre of origin)
Trifolium	*alexandrinum*	Syria	Egypt
Medicago	*sativa*	Asia Minor	Near East and central Asia

Table 1. Centre of origin of different tropical forage legumes.

crops [17]. Among the international organizations major forage germplasm repositories are International Livestock Research Institute (ILRI), Nairobi, CIAT Columbia; ICARDA Syria; CSIRO-Australia, IGER-UK, USDA-Fort Collins. Forage germplasm diversity in these organizations is part of a Consultative Group of International Agricultural Research (CGIAR) coordinated activity in plant genetic resources. The ILRI Gene bank conserves more than 18 thousand accessions of forages from over 1000 species. This is one of the most diverse collections of forage grasses, legumes and fodder tree species held in any gene bank in the world [18]. CIAT gene bank keeps 35,898 accessions of beans, for 44 species of the genus *Phaseolus* from 109 countries, and 23,139 forage accessions belonging to 668 different species of grasses and legumes from 72 countries, that have been introduced over the past 30 years [19]. The IITA gene bank holds the world's largest and most diverse collection of cowpeas, with 15,122 unique samples from 88 countries, representing 70% of African cultivars and nearly half of the global diversity.

Indian sub-continent being one of the world's mega centres of crop origin and crop plant diversity, represents a wide spectrum of eco-climate and reported diversity of 21 forage legumes genera *viz.*, *Desmodium, Lablab, Stylosanthes, Vigna, Macroptelium, Centrosema* and browse plants including *Leucaena, Sesbania, Albizia, Bauhinia, Cassia, Grewia,* etc. (**Table 2**). Diversity of cultivated and range legumes were collected in form of 3261 diverse germplasm accessions through different indigenous and exotic germplasm collection programme. Collected diversity of forage legumes were evaluated and sources for different biotic and abiotic stress tolerance were identified apart from >50 cultivars in different forage legumes for different geographic regions developed. Crop wild relatives (CWR) being the reservoirs of genes for stress tolerance and quality have been utilized for genetic enhancement of forage legumes. The main centre of diversity for tropical legumes *viz.*, *Dolichos, Desmodium, Vigna* and *Crotalaria* is peninsular India and subtropical legumes *viz. Teramnus, Atylosia, Pueraria* and *Mucuna* are mainly confined to north eastern region. Likewise, rich genetic wealth for the temperate legumes namely *Medicago, Melilotus, Trifolium* and *Hedysarum* is distributed in western Himalayan region [20]. Besides, India possesses enormous diversity of minor and under-utilized fodder species such as *Agrostis alba, Desmodium parvifolium, Leptochloa fusca, Potentilla fruticosa, Rhynchosia minima* and *Salvadora persica* [21]. The forage genetic wealth of India distributed in 15 agro-climatic zones has been summarized in **Table 2**.

The National Bureau of Plant Genetic Resources (NBPGR) is the nodal agency for characterization, evaluation, maintenance, conservation, documentation and distribution of germplasm resources in India. Currently a total of 4594 accessions of different forage crops including cereal forages (1167), grasses (11,160, range legumes (1443), forage millets (781) and others [85] are being maintained at long term storage (LTS) module of National Gene Bank at NBPGR, New Delhi [22]. Indian Grassland and Fodder Research Institute (IGFRI) is a unique R&D organization in South Asia for sustainable agriculture through quality forage production for improved animal productivity. IGFRI being the National Active Germplasm Sites (NAGS) on forages works with its three regional stations and All India Coordinated Research Project (AICRP) on forage crops with 18 coordinated centres. At present IGFRI maintains more than 8000 accessions of 19 major forage crops including cereal forages, forage legumes, grasses and fodder tree at midterm storage [23].

http://dx.doi.org/10.5772/intechopen.81186

S. no.	Agro climatic zone/regions	Subzones/sub regions	Prominent forage genetic resources
1	Western Himalayan Region	Jammu & Kashmir, Himachal Pradesh, Uttarakhand Hills	*Medicago* spp., *Arundinella nepalensis,* Chrysopogon, *Dactylis glomerata, Eleusine, Echinochloa, Festuca, Zea mays,* Kikui *grass*
2	Eastern Himalayan Region	Sikkim, Arunachal Pradesh, Meghalaya, Nagaland, Manipur, Tripura, Mizoram, Assam, Jalpaiguri and Cooch Bihar district of West Bengal	Rice bean, maize, range grasses, *Brachiaria,* broom grass and lablab bean
3	Lower Gangetic Plains	Basin plains, central alluvial plains, alluvial coastal plains and *Rarh* plains	Rice bean, guinea grass, coix and range grasses
4	Middle Gangetic Plains	12 districts of eastern Uttar Pradesh and 27 districts of Bihar plains	Maize, cowpea, rice bean, *Pennisetum pedicellatum* and coix.
5	Upper Gangetic Plains	central, south-western and northern-western Uttar Pradesh	Maize, sorghum, cowpea Senji, *Dichanthium,* sehima and *Heteropogon*
6	Trans-Gangetic Plains	Punjab, Haryana, Delhi, Chandigarh and Sri Ganganagar district of Rajasthan	Guar, maize, bajra, berseem, lucerne, guinea grass, sorghum and cowpea
7	Eastern Plateau and Hills	(i) Sub region of Wainganga, Madhya Pradesh, eastern hills and Orissa inland; (ii)Orissa northern, Madhya Pradesh, eastern hills and plateau; (iii) north and eastern Chota Nagpur hills and plateau; (iv) Chota Nagpur south, West Bengal hills and plateau, and (v) Chhattisgarh and south-western Orissa hills.	Cowpea, rice bean, *Pennisetum pedicellatum,* guinea grass, *Dichanthium* spp. and *Atylosia*
8	Central Plateau Hills	46 districts of Uttar Pradesh, Madhya Pradesh and Rajasthan	Maize, cowpea, rice bean, *P. pedicellatum, Coix, Atylosia,* sorghum, bajra, guar, *Cenchrus,* range grasses and legumes
9	Western Plateau and Hills	Maharashtra, parts of Madhya Pradesh and one district of Rajasthan	Maize, sorghum, *Dichanthium spp.* pearl millet, *Dichanthium carzacosum, Vicia,* cowpea, rice bean, Cenchrus, range grasses and legumes
10	Southern Plateau and Hills	35 districts of Andhra Pradesh, Karnataka and Tamil Nadu	small millet, *Heteropogon, Dichanthium* sehima and *Stylosanthes* sp.
11	East Coast Plains and Hills	(i) Coastal Orissa (ii) North-Coastal Gujarat (iii) South-Coastal Andhra Pradesh, North-Coastal Tamil Nadu (v) Thanjavur and (vi) South Coastal Tamil Nadu.	cowpea, rice bean, guinea grass, coix, small millet, sorghum, *Heteropogon, Dichanthium* and *Stylosanthes* sp.
12	West Coast Plains and Hills	Western coast of Tamil Nadu, Kerala, Karnataka, Maharashtra and Goa	Congo, signal grass, *Paspalum, panicum, Digitaria, Brachiaria, Iseilema laxum, Isilemia* and *Vicia*
13	Gujarat Plains and Hills	19 districts of Gujarat	Lucerne, sorghum, small millet, pearl millet, chioori, range grasses and legumes
14	Western Dry Region	Nine districts of Rajasthan	Guar, moth, cowpea, sorghum, pearl millet and *Cenchrus* spp.
15	Island Region	Territories of the Andaman and Nicobar Islands and Lakshadweep	

Adopted from Singh et al. [77].

Table 2. List of prominent forage genetic resources distributed in 15 agro climatic zones of India.

5. Problems associated with breeding of tropical forage legumes

Tropical forage legumes breeding programmes are associated with certain unique problems. Most of the tropical pasture legumes still possess traits of wild plants that include seed shattering, small seed size, seed dormancy, relatively slow germination rates, etc. In most of the cases we have very little knowledge about the basic biology of the species. Some of the problems include overlapping of vegetative and reproductive growth phases, uneven pod setting, non-synchronous maturity and seed shattering in forage legumes [24]. Inherent heterozygosity as most forage species are cross pollinated. Self-incompatibility limits the extent to which they may be inbred; small floral parts make artificial hybridization tedious; poor seed producers; or produce seed with low viability as well as inherently low seedling vigor and competitive ability. Many forage species produce weak seedlings and stands are not easily established. Strains may perform differently with different systems of grazing management. Persistence of perennial tropical forage legumes is not as a single trait, but rather as a complex of traits dependent on various factors, such as diseases, insects, abiotic stresses, or management stress. Fertility barriers of one sort or another are very common in tropical forage legume breeding *viz.*, berseem [25], owing to the wild nature of the species and inadequate knowledge of inter- or intra-specific variation.

6. Major forage legumes of India

6.1. Egyptian clover (*Trifolium alexandrinum* L.)

The genus *Trifolium* from the tribe *Trifolieae* of the family Leguminosae (Fabaceae) is important for its agricultural value. A few of the 237 species of this large genus have actually been cultivated [26], out of which 25 species are important as cultivated and pasture crops [27]. Egyptian clover or berseem (*T. alexandrinum* 2n = 16) is commonly cultivated as winter annual in the tropical and subtropical regions. Berseem is popular due to its multicut [4–8] nature, providing fodder for a long duration (November to May), very high quantum of green fodder (85 t/ha) and better quality of fodder (20% crude protein), high digestibility (up to 65%) and palatability. Berseem was introduced in India from Egypt in 1904, and has been established as one of the best *Rabi* (winter season) fodder crop in entire North West Zone, Hill Zone and part of Central and Eastern Zone of the country, occupying more than two million hectare [28].

Berseem being an introduced crop in India, the most important drawback in genetic improvement has been the lack of genetic variability [29, 30]. Variability in the existing gene pool has been induced through mutation, polyploidization and inter-specific hybridization. High biomass production potential along with extended growth period and resistance to biotic stresses specially root rot and stem rot have been the main target traits that were to be improved genetically. Different genetic improvement programmes carried out in various research institutes/universities by utilizing breeding approaches like selection, polyploidy and mutation resulted in the development of >15 varieties for different berseem growing regions of India. Inter-specific hybridization have been used to improve resistance to biotic and abiotic stresses and extended length of the vegetative period because genes for wide scale adaptability are widely distributed in several wild species of *Trifolium* (**Table 3**). Interspecific hybrids of berseem with *Trifolium apertum* [31], *T. constantinopolitanum* [32], *T. resupinatum* [33] and *T. vesiculosum* [34] were successfully developed and progenies of interspecific hybrids showed introgression of various desirable traits, including late flowering and resistance to root rot and stem rot diseases.

A major breakthrough in berseem breeding in India was achieved through induction of polyploidy. The work on polyploidization of berseem genome was started with the aim to induce greater leaf and stem size [35, 36]. Autotetraploid induced by using colchicine treatment, and selection at tetraploid level resulted in the development of first polyploid variety 'Pusa Giant' with more fodder production and good regeneration capacity, uniform and higher yield throughout the season than diploid varieties released for general cultivation in India [37]. Another big achievement in polyploidy breeding was achieved at IGFRI, Jhansi by developing an autotetraploid variety namely 'Bundel Berseem-3' through colchiploidy followed by recurrent single plant selection followed with mass selection [28]. Major success in Berseem breeding was achieved by induction of longer duration mutant in Mescavi variety through gamma ray treatment which resulted in 'BL-22' a variety released

Species	Chromosome number (2n)	Desirable characters	References
T. alexandrinum ecotype Mescavi	2n = 16	Annual, multicut, highly productive, crude protein, high digestibility and palatability, basal branching	[31]
T. alexandrinum ecotype Fahli	2n = 16	Annual, single cut, self-compatible, stem branching	[78]
T. alexandrinum ecotype Saidi	2n = 16	Annual, 2–3 cut, stem and basal branching	[78]
T. berytheum	2n = 16	Biotic resistance	[79]
T. salmoneum	2n = 16	Biotic resistance	[79]
T. apertum	2n = 16	Annual, profuse basal branching, late flowering, resistance against root rot and stem rot, high protein content	[31, 79]
T. meironense	2n = 16	Biotic resistance	[31]
T. resupinatum	2n = 16	Root rot and stem rot resistance, soil alkalinity tolerance	[33, 80]
T. constantinopolitanum	2n = 16	Profuse basal branching, resistance against root rot and stem rot	[32]
T. vesiculosum	2n = 16	Lateness, disease resistance	[25]

Table 3. Desirable characters in berseem ecotypes and wild *Trifolium* species.

in 1988 for temperate and north west zone; and 'BL-180' released in 2006 for cultivation in north-west zone of India [28]. Protocol for in vitro plant regeneration from meristematic tissue and the establishment of regenerable callus culture have been developed in Berseem and related species viz., *Trifolium glomeratum, T. apertum, T. resupinatum* [38–40]. Embryo rescue technique has been effectively utilized to overcome the problems of post fertilization barriers in interspecific crosses of berseem with *Trifolium apertum, T. constantinopolitanum, T. resupinatum* and *T. vesiculosum* [31–34]. Recently, SSR based markers were developed for large scale utilization programme in Berseem [30]. Few studies on genetic diversity in Berseem and related *Trifolium* species were reported by using isozymes [29] and molecular markers [41].

6.2. Stylosanthes

The genus *Stylosanthes* comprises approximately 40 species, distributed in the tropical [42], subtropical and temperate regions areas of America, Africa, and Southeast Asia. It can be grouped into two subgeneric sections, Stylosanthes and Stylosanthes. Most species are diploid (2n = 20) but polyploid species (2n = 40 and 2n = 60) also exist. Six species, namely *Stylosanthes scabra, S. seabrana, S. hamata, S. guianensis, S. humilis* and *S. viscosa*, are predominantly used as fodder legume in humid to semi-arid tropics of India (**Table 4**). These are very popular and have been widely adapted due to their ability to restore soil fertility, improve soil physical properties, and provide permanent vegetation cover as well as to provide nutritious fodder. The most specific problems associated with Stylosanthes are the limited variations of available germplasm and the susceptibility to anthracnose disease caused by the fungus *Colletotrichum gloeosporioides*. In the past, mainly five species of Stylosanthes

Species	Chromosome	Specific features
S. scabra	2n = 4x = 40	Adapted in low rainfall areas (325 mm rainfall), suitable for semi-arid areas of Maharashtra, Andhra Pradesh, Karnataka and Tamil Nadu, *S. seabrana* and *S. viscosa* are known progenitor of *S. scabra*
S. hamata	2n = 2x = 20 2n = 4x = 40	Diploid *S. hamata* and *S. humilis* are the two progenitors of this species (Curtis et al., 1995), highly palatable, grazing tolerant
S. viscosa	2n = 2x = 20	Early emergence and highly stickiness of the leaves and stems, drought tolerant, grows on poor soils, some resistance to anthracnose, acaricidal properties
S. humilis	2n = 2x = 20	Tolerance for salinity, susceptible to anthracnose, hairs on stems and leaves are some of the important features helpful in identifying the species
S. guianensis	2n = 2x = 20	Suitable for humid and higher rainfall regions, adapted to acid infertile soils, tolerant of Al and Mn
S. fruticosa	2n = 4x = 40	Allotetraploid, drought tolerant

Table 4. Important *Stylosanthes* spp. with specific features.

(*S. hamata, S. scabra, S. humilis, S. viscosa* and *S. guianensis*) have been introduced primarily from Australia and evaluated at different sites in India [43–45]. This was in addition to the native perennial *S. fruticosa* Alston, which is widely distributed throughout the southern peninsular regions [46].

Testing and evaluation of wide germplasms carried out at IGFRI on acid and saline soil which contribute major part of the soils of India, indicated better adaptation of *S. hamata* and *S. seabrana* lines over other species in salinity. The potential of S. *seabrana* for tropical and subtropical regions of the country with clay and heavy soils, cool winters and distinct wet-dry seasonal conditions directed the use of this species in developing new breeding approach. The one could be based on the finding that it is the second progenitor of *S. scabra* which in turn elucidated the evolution of one of the most important *Stylosanthes* species, *S. scabra* may lead to important impacts on the efforts of improving *S. scabra* [47]. It may be possible to artificially synthesize *S. scabra* using pre-selected *S. viscosa* and *S. seabrana* accessions [48]. These artificial *S. scabra* genotypes could be used directly or more likely, be used in breeding programs. By doing so the genetic variation existing in the two diploid progenitor species would become available in improving the allotetraploid *S. scabra*. So far developed map and linked markers with anthracnose resistance also provide the opportunity to use them after converting them in sequence tagged sites (STS) or sequence characterized amplified region (SCAR) and then using them in direct breeding programs.

6.3. Alfalfa (*Medicago sativa* L.)

Genus *Medicago* is one of the oldest forage legume comprising 60 perennial and 35 annual species, distributed mainly around the Mediterranean basin, cultivated throughout the world in diverse environments ranging both temperate and tropical environments [49]. It is generally agreed that the basic chromosome number for the genus *Medicago* are x = 7 and x = 8. Its ploidy varies from diploid (2n = 16) to polyploid (2n = 32, 48, 64). Perennial species are mainly tetraploids (2n = 4x = 32) and allogamous, however diploid (2n = 2x = 16) and hexaploid (2n = 6x = 48) cytotypes have also been reported [50]. *Medicago sativa* (alfalfa or lucerne) is widely cultivated as the most important forage legume in the temperate areas of the world. Lucerne is native to South West Asia as indicated by occurrence of wild types in the Cancasus and in mountainous region of Afghanistan, Iran. *M. sativa* complex, comprises of several members at the same ploidy level e.g., *M. falcata, M. media* and *M. glutinosa,* which freely intercross, without any hybrid sterility in the F_1 or later generations [51]. In India, it is grown in Maharashtra, Gujarat, Andhra Pradesh, Karnataka, Tamil Nadu, Haryana, Madhya Pradesh, Rajasthan, Punjab. The major breeding objectives in the crop include vigorous tall growing plants, better branching, quick regeneration, and balance between seed and forage yield and persistence.

Genetic resources for alfalfa improvement are limited and restricted to the *M. sativa* complex but tolerant sources for biotic and abiotic constraints are lacking in the complex [52].

The annual and perennial species of the genus *Medicago* are the reservoir of several useful agronomic traits, including disease and insect resistance and potential salt and drought tolerance having direct implication in cultivated alfalfa improvement (**Table 5**). Most of the lucerne cultivars grown in the country and worldwide are susceptible to many diseases and insect pests and the most serious constraint is the alfalfa weevil (*Hypera postica* Gyll.) [53]. Resistance to weevil has been reported in several annual species such as *M. scutellata, M. prostrata, M. turbinata* and *M. intertexta* [54–57]. Genes conferring resistance to aphid have been identified in *M. rugosa, M. scutellata* and *M. littoralis* [58]. Similarly, three woody species *viz. M. arborea, M. strasseri* and *M. citrine* of the section *Dendrotelis* have been reported as excellent sources for incorporating drought and salt tolerance in *M. sativa* [59–61]. However, due to post fertilization barrier, interspecific hybridization is difficult, so we may need to use biotechnological tools like ovule-embryo culture and electroporation.

Inter specific hybrids of *M. sativa* with some of the perennial species viz. *M. cancellata, M. glomerata, M. papillosa, M. prostrata, M. rhodopea* and *M. saxatilis* have been recovered by conventional crosses [51]. However, pollen and embryological studies demonstrated that there exist strong post fertilization barriers for recovering hybrids between *M. sativa* and annual species [62]. Utilizing embryo culture and fertilized pod culture techniques interspecific hybrids were obtained between *M. sativa* and many other annual species however, no hybrids were produced between *M. sativa* and weevil resistant *M. scutellata* [63, 64]. Bauchan and Elgin [65] reported chromosomal incompatibility and presence of two SAT chromosomes in *M. scutellata* as the major barriers for getting interspecific hybrids between *M. sativa* and *M. scutellata*. Utilizing protoplast fusion technique S_1 plants were obtained between *M. sativa* and *M. rugosa* and it was confirmed by genomic *in situ* hybridization (GISH) that small portions of *M. rugosa* chromosomes were present in the hybrid however, it is not clear that in which chromosome the resistance genes are present [50].

A lot of molecular information has been generated across species. However, information from *M. truncatula* on marker-trait association is unlikely to be exploitable in lucerne, considering the large differences between annual and perennial [66]; in addition to the differences due to the ploidy level which may further contribute to the inconsistent genetic control of some morpho-physiological traits between the two species [67]. Some breeding goals such as region-specific adaptation; drought-tolerance; improvement for forage quality should be considered [68]. Attempts have been made to produce transgenic alfalfa containing fungal chitinase gene for resistance against fungal pathogens [69], tolerance to abiotic stresses such as salt and cold [70, 71], improved forage quality [72], and sulfur-containing amino acids [73], value addition by making it an edible forage vaccine [74]. In recent years the breeding strategies for Lucerne are more towards utilizing potential of polycross methods followed with phenotypic selection. It has resulted in development of a few cultivars in recent years. The future strategies should include development of cold and drought hardy lucerne with degree of persistence for pasture and meadows, increasing genetic base, high seed production, stress tolerance, diseases and pest resistance etc.

Species	Annual/ perennial	Chromosome number (2n)	Distribution	Desirable traits	References
M. dzhawakhetica Bordz.	Perennial	32	Western Mediterranean region	Cold tolerance and resistance to *Phoma medicaginis*	[81]
M. suffruticosa Ram.	Perennial	—	—	Resistance to *Phoma medicaginis*, deep taproot system and high palatability	[81]
M. cancellata M.B.	Perennial	48	Russia	Resistance to *Stemphyllium* leaf spot	[82]
M. prostrata Jacq.	Perennial			Resistance to alfalfa weevil and potato leafhopper	[54]
M. scutellata (L.) Miller	Annual	30	Mediterranean Basin, Southern Ukraine	High biomass production, Resistance to alfalfa weevil and aphid	[83]
M. turbinata (L.) All.	Annual	—	Mediterranean Basin	Resistance to alfalfa weevil	[54, 56]
M. intertexta (L.) Miller	Annual	16	West Mediterranean Basin	Resistance to alfalfa weevil	[54, 57]
M. rugosa Desr.	Annual	30	Mediterranean Basin	Resistance to aphid	[58]
M. littoralis Rohde ex Lois	Annual	—	Mediterranean Basin, East Europe, Caucasus	Resistance to aphid	[58, 83]
M. polymorpha L	Annual	14	Europe, North Africa, Middle East, Ukraine, Georgia, Central Asia	Plant height, high seed production potential	[83]
M. lupulina L	Annual			Excellent species for sustainable agriculture, reported to improve soil health, reduce diseases and save moisture	
M. arborea Hutch.	Perennial	32	Mediterranean region	Woody species, ornamental value, drought and salt tolerant	[59, 61]
M. strasseri Greuter et al.	Perennial	32	Crete Iceland	Woody species, drought and salt tolerance	[60]
Medicago citrine (Font Quer) Greuter	Perennial	48	Balearic Islands	Highly drought and salt tolerant species within the section *Dendrotelis*	[84, 85]
M. truncatula Gaertner	Annual	16	Mediterranean Basin, East Europe, Russia	Genes possessing broad spectrum resistance to anthracnose, stay green genes	[86]

Table 5. Annual and perennial *Medicago* species and their desirable characters.

6.4. Cowpea (*Vigna unguiculata* (L.) Walpers)

Cowpea (2n = 2x = 22, genome size = 620 Mb) also known as 'black eye pea' or 'hungry-season crop' is an annual food and forage crop mostly grown throughout the semi-arid tropics in parts of Asia, Africa, Southern Europe, Southern United States, and Central and South America (Singh 2005). It can be grown throughout the year due to its short duration and fast growing nature. It is suitable for inter, mixed and relay cropping system. Cultivated cowpea, which is in subspecies *unguiculata*, is divided into five cultivar groups namely *Unguiculata*, *Sesquipedalis* (yard-long-bean), *Textilis*, *Biflora* and *Melanophthalmus* [75]. The commonly cultivated cowpea belongs to cultivar group *Unguiculata* the most widespread and economically important group of the species. They are pulse and vegetable and forage types. Other cultivar group *Biflora* also known as 'catjang cowpea' mainly cultivated in South Asia (India, Sri Lanka) as a pulse or as forage for hay and silage, and as a green manure crop. In Australia and Asia cowpea is primarily a fodder crop, but is also used for green manure or as a cover crop [76]. In India, the crop is cultivated around 6.5 lakh ha with 3 lakh as fodder crop in Rajasthan, Gujarat, Maharashtra, Karnataka and Tamil Nadu [24].

Cowpea was first introduced to India 1000–1500 years ago and now Indian-subcontinent appears to be a secondary centre of diversity. In India a large numbers of varieties for vegetable, pulse and fodder purpose have been developed. The breeding objectives have focused around developing lines with terminal drought tolerance, early maturity, erect growth to fit in cropping systems and enabling improved radiation use efficiency, high harvest index and resistance to diseases. The desirable traits in forage cowpea varieties are leafiness with indeterminate growth to get green fodder for a longer period. International Institute of Tropical Agriculture (IITA) has developed several dual purpose cultivars of cowpea with high grain and biomass yields and erects habit for intercropping/mixed farming purposes. In future development of cowpea lines against various forms of root-knot nematode, cowpea aphids and *Fusarium* wilt, is required. Further, development of transgenic cowpea lines with resistance to major insect pests can also be a breakthrough in cowpea breeding.

7. Conclusion

Tropical forage legumes were promoted in the past with the major focus on livestock production in India. This has led to a substantial decrease in research on tropical forage legumes. In view of current climate change problems and environmental concerns, research on forage legumes should be resumed with adequate funding support at national and international levels. Newer biotic and abiotic stress tolerant varieties should be developed for the changing environmental conditions. Forage legumes have potential to contribute significantly to environment-friendly agricultural land use and sustainable livestock production in the tropics.

Author details

Tejveer Singh, Srinivasan Ramakrishnan*, Sanat Kumar Mahanta, Vikas C. Tyagi and Ajoy Kumar Roy

*Address all correspondence to: srinivasmic@gmail.com

ICAR-Indian Grassland and Fodder Research Institute, Jhansi, India

References

[1] BAHS. Basic Animal Husbandry Statistics. New Delhi: Department of Animal Husbandry, Dairying and Fisheries, Ministry of Agriculture, Government of India; 2012. p. 130

[2] Ghosh PK, Mahanta SK. Forage resource development in India: Looking ahead. In: Agriculture Year Book. New Delhi, India: Agriculture Today; 2014. pp. 134-140

[3] NAAS. Augmenting Forage Resources in Rural India: Policy Issues and Strategies. Policy Paper No. 80. New Delhi: National Academy of Agricultural Sciences; 2016. pp. 1-16

[4] Mahanta SK, Singh KK, Das MM, Das N. Forage based feeding of livestock. In: Das N, Misra AK, Maity SB, Singh KK, Das MM, editors. Forage for Sustainable Livestock Production. Delhi: Satish Serial Publishing House; 2009. pp. 407-426

[5] Peoples MB, Ladha JK. Herridge DF enhancing legume N_2 fixation through plant and soil management. Plant and Soil. 1995;**174**:83-101

[6] Ladha JK, Kundu DK, Angelo V, Coppenolle MG, Peoples MB, VR C, et al. Legume productivity and soil nitrogen dynamics in lowland rice-based cropping systems. Soil Science Society of America Journal. 1996;**60**:183-192

[7] Smil V. Nitrogen in crop production: An account of global flows. Global Biogeochem Cycles. 1999;**13**:647-662

[8] Rao DLN, Balachandar D. Nitrogen inputs from biological nitrogen fixation in Indian agriculture. In: Abrol et al, editors. The Indian Nitrogen Assessment: Sources of Reactive Nitrogen, Environmental and Climate Effects, Management Options, and Policies. Elsevier Inc. 2017. pp. 117-132. DOI: 10.1016/b978-0-12-811836-8.00008-2

[9] Herridge DF, Peoples MB, Boddey RM. Global inputs of biological nitrogen fixation in agricultural systems. Plant and Soil. 2008;**311**:1-18

[10] Yasmin K, Cadisch G, Baggs EM. Comparing ^{15}N labelling techniques for enriching above- and below ground components of the plant–soil system. Soil Biology and Biochemistry. 2006;**38**:397-400

[11] Mahieu S, Fustec J, Faure M-L, Corre-Hellou G, Crozat Y. Comparison of two 15N labelling methods for assessing nitrogen rhizodeposition of pea. Plant and Soil. 2007;**295**: 193-205

[12] Boddey RM, Carvalho INO, Rezende CP, Cantarutti RB, Pereira JM, Macedo R, et al. The benefit and contribution of legumes and biological N_2 fixation to productivity and sustainability of mixed pastures. In: Evangelista AR, CLS A, Casagrande DR, Lara MAS, Bernardes TF, editors. Proceedings of the 1st International Conference on 'Forages in Warm Climates'. Lavras, MG, Brazil: Universidade Federal de Lavras; 2015. pp. 103-140

[13] Lesturguez G, Poss R, Hartmann C, Bourdon E, Noble A, Ratana-Anupap S. Roots of *Stylosanthes hamata* create macropores in the compact layer of a sandy soil. Plant and Soil. 2004;**260**:101-109

[14] Schultze-Kraft R, Rao IM, Peters M, Clements RJ, Bai C, Liu G. Tropical forage legumes for environmental benefits: An overview. Tropical Grasslands-Forrajes Tropicales. 2018;**6**:1-14

[15] Trivedi BK. Grasses and Legumes for Tropical Pasture. Jhansi: Indian Grassland and Fodder Research Institute; 2010

[16] http://apps3.fao.org/wiews/

[17] FAO. The State of *ex situ* Conservation: The Second Report on the State of the world's PGRFA. 2010. p. 351. http://www.fao.org/docrep/013/i1500e/i1500e03.pdf

[18] www.ilri.org

[19] www.ciat.cgiar.org

[20] Arora RK, Chandel KPS. Botanical source areas of wild herbage legumes in India. Tropical Grasslands. 1972;**6**:213-221

[21] Bhag M. Neglected and underutilized crop genetic resources for sustainable agriculture. Indian Journal of Plant Genetic Resources. 2007;**20**:1-14

[22] Joshi V, Tyagi V, Kak A, Lal A. Status of genetic resources of forage crops in India: A review. Indian Journal of Plant Genetic Resources. 2009;**22**(3):243-252

[23] Sahay G, Sevanayak D, Tyagi VC, Bhardwaj N, Shmed S. Tropical and subtropical forage genetic resources of India: Their conservation and utilization. Forage for Future. 2018;**6**:3

[24] Roy AK, Malaviya DR, Kaushal P. Genetic improvement of fodder legumes especially dual purpose pulses. Indian Journal of Genetics and Plant Breeding. 2016;**76**:608-625

[25] Malaviya DR, Roy AK, Kaushal P, Bhaskar RB, Kumar B. Evaluation of *Trifolium* species for defining multiple use gene pool for tropical *Trifolium* species. Indian Journal of Genetics. 2004;**64**(3):251-252

[26] Zohary M, Heller D. The Genus *Trifolium*. Jerusalem: The Israel Academy of Sciences and Humanities; 1984

[27] Lange O, Schifino-Wittmann MT. Isozyme vtriation in wild and cultivated species of the genus *Trifolium* L. (Leguminosae). Annals of Botany. 2000;**86**:339-345

[28] Pandey KC, Roy AK. Forage Crops Varieties. Jhansi, India: IGFRI; 2011

[29] Malaviya DR, Kumar B, Roy AK, Kaushal P, Tiwari A. Estimation of variability of five enzyme systems among wild and cultivated species of *Trifolium*. Genetic Resources and Crop Evolution. 2005;**52**:967-976

[30] Verma P, Chandra A, Roy AK, Malaviya DR, Kaushal P, Pandey D, et al. Characterization and cross-species transferability of genomic SSR markers in berseem (*Trifolium alexandrinum* L.), an important multi-cut annual forage legume. Molecular Breeding. 2015;**35**:23

[31] Malaviya DR, Roy AK, Kaushal P, Kumar B, Tiwari A, Lorenzoni C. Development and characterization of interspecific hybrids of *Trifolium alexandrinum* × *T. apertum* using embryo rescue. Plant Breeding. 2004;**123**:536-542

[32] Roy AK, Malaviya DR, Kaushal P, Kumar B, Tiwari A. Inter-specific hybridization of *Trifolium alexandrinum* with *T. constantinopolitanum* using embryo rescue. Plant Cell Reports. 2004;**22**:705-710

[33] Kaushal P, Malaviya DR, Roy AK, Tiwari A, Kumar B. *Trifolium alexandrinum* × *T. resupinatum*—Interspecific hybrids developed through embryo rescue. Plant Cell, Tissue and Organ Culture. 2005;**83**(2):137-144

[34] Kaur A, Kaur KP, Kalia A. Generation of inter-specific hybrids between *Trifolium vesiculosum* and T. *alexandrinum* using embryo rescue. Euphytica. 2017;**213**:253

[35] Mehta R, Swaminathan M. Studies on induced polyploids in forage crops. 1. Survey of previous work. Indian Journal of Genetics. 1957;**17**:27-57

[36] Sikka SM, Mehta RK, Swaminathan MS. Studies on induced polyploids in forage crops II. Colchicine treatment methods for berseem and senji. Indian Journal of Genetics. 1959;**19**(1):90-97

[37] Mehta R, Swaminathan M. 'Pusa Giant' berseem. Indian Farming. 1965;**15**(6):4-5

[38] Kaushal P, Malaviya DR, Roy AK, Kumar B. In vitro response of immature zygotic embryos of *Trifolium alexandrinum* (Egyptian clover). Plant Cell Biotechnology and Molecular Biology. 2004;**5**(3-4):109-114

[39] Kaushal P, Tiwari A, Roy AK, Malaviya DR, Kumar B. *In vitro* regeneration of *Trifolium glomeratum*. Biologia Plantarum. 2006;**50**(4):693-696

[40] Malaviya DR, Roy AK, Tiwari A, Kaushal P, Kumar B. *In vitro* callusing and regeneration in *Trifolium resupinatum*—A fodder legume. Cytologia. 2006;**71**(3):229-235

[41] Kaila V, Sood VK, Chaudhary HK, Bhandari JC, Sood A, Mittal RK. Morphological and RAPD markers-mediated assessment of genetic diversity amongst various *Trifolium* species and identification of potential ideotypes for genetic upgradation of berseem under changed climate in mid-hills of north-west Himalayas. Indian Journal of Genetics. 2009;**69**(4):1-6

[42] Kirkbide JH, De Kirkbride CG. Typification of *Stylosanthes* (Leguminosae) and its sections. Taxon. 1985;**36**:455-458

[43] Rai P, Patil BD. Production potential of *Stylosanthes* in pure and mixed swards. Indian Journal of Range Management and Agroforestry. 1985;**6**:35-57

[44] Chatterjee BN, Singh RD, Matti S. An agronomical appraisal of *Stylosanthes* in eastern region of India. Indian Journal of Range Management and Agroforestry. 1985;**6**:27-33

[45] Ramesh CR, Mal B, Hazara CR, Sukanya DH, Ramamurthy V, Chakraborty S. Status of *Stylosanthes* development in other countries. III. *Stylosanthes* development and utilization in India. Tropical Grasslands. 1997;**31**:467-475

[46] Hooker JD. Flora of British India. In: Sabiaceae to Cornaceae. London: L. Reeve and Co. Ltd; 1879. p. 148

[47] Tewari S, Chandra A. Genetical assessment of diploid progenitors of *S. scabra* by isozyme, RAPD and STS markers: A possible strategy for improvement of drought tolerant allo-tetraploid *S. scabra* species. Euphytica. 2008;**162**(1):39-50

[48] Liu CJ, Musial JM. *Stylosanthes* sp. aff. *S. scabra*: A putative diploid progenitor of *Stylosanthes* scabra (Fabaceae). Plant Systematics and Evolution. 1997;**208**:99-105

[49] Small E, Jomphe M. A synopsis of the genus Medicago. Canadian Journal of Botany. 1989;**67**:3260-3294

[50] Mizukami Y, Kato M, Takamizo T, Kanbe M, Inami S, Hattori K. Interspecific hybrids between *Medicago sativa* L. and annual *Medicago* containing Alfalfa weevil resistance. Plant Cell Tissue and Organ Culture. 2006;**84**:79-88

[51] Lesins KA, Lesins I. Genus *Medicago* (Leguminosae) a Taxonomic Study. The Hague, The Netherlands: Junk bv Publishers; 1979. p. 228

[52] Barnes DK, Bingham ET, Murphy RP, Hunt OJ, Beard DF, Skrdla WH, et al. Alfalfa germplasm in the United States: Genetic vulnerability, use, and maintenance. In: USDA-ARS Tech. Bull. Hyattsville, MD: USDA-ARS; 1977. p. 1571

[53] Chandra A, Pandey KC. Assessment of genetic variation in lucerne (*Medicago sativa* L.) using protease inhibitor activities and RAPD markers. Journal of Environmental Biology. 2011;**32**:559-565

[54] Sorensen EL, Horber EK. Glandular trichomes on Medicago: Building self defense against insects. In: Rep. of the Twenty-Ninth Alfalfa Improvement Conf. 1984. p. 135

[55] Kleitner GL, Sorensen EL. Erect glandular trichomes of *Medicago scutellata* (L.) Mill: Gland deveropment and early secretion. Botanical Gazette. 1983;**144**(2):165-174

[56] Chandra A, Pandey KC, Singh UP. *Medicago scutellata*—A possible source of weevil resistance for lucerne improvement. Indian Journal of Plant Genetic Resources. 2006; **19**:291-293

[57] Chandra A, Pandey KC. Effect of proteinase inhibitors on Indian alfalfa weevil (Hypera postica Gyll.) growth and development. Acta Physiologiae Plantarum. 2008;**30**:501-505

[58] Verma JS, Mishra SN. Advances in forage plant improvement in upper Gangetic Plains. In: Hazra CR, Mishri B, editors. New Vistas in Forage Production. Jhansi-284003, India: Indian Grassland and Fodder Research Institute. 1995. pp. 83-96

[59] Olives G. La alfalfa arbórea. In: Pesca y Alimentación. Madrid: Ministerio de Agricultura; 1969

[60] Greuter W, Matthäs U, Risse H. Notes on Cardaegan plants. 3. Medicago strasseri, a new leguminous shrub from Kriti. Wild. 1982;**12**:201-206

[61] Rosato M, Castro M, Rosselló JA. Relationships of the Woody *Medicago* species (section *Dendrotelis*) assessed by molecular cytogenetic analyses. Annals of Botany. 2008;**102**(1):15-22

[62] MCCOY T. Interspecific hybridization of *Medicago sativa* L. and *M. rupestris* M.B. using ovule-embryo-culture. Canadian Journal of Genetics and Cytology. 1985;**27**(2):238-245

[63] Wang JW, Sorensen EL, Liang GH. In vitro culture of pods from annual and perennial *Medicago species*. Plant Cell Reports. 1984;**3**:146-148

[64] McCoy TJ, Smith LY. Interspecifi c hybridization of perennial *Medicago* species using ovule-embryo culture. Theoretical and Applied Genetics. 1986;**71**:772-783

[65] Bauchan GR, Elgin JH Jr. A new chromosome number for the genus *Medicago*. Crop Science. 1984;**24**:193-195

[66] Volenec JJ, Cunningham SM, Haagenson DM, Berg WK, BC J, Wiersma DW. Physiological genetics improvement: Past failures, future prospects. Field Crops Research. 2002;**75**:97-110

[67] Bingham ET, Groose RW, Woodfield DR, Kidwell KK. Complementary gene interactions in alfalfa are greater in autotetraploids than diploids. Crop Science. 1994;**34**:823-829

[68] Annicchiarico P, Scotti C, Carelli M, Pecetti L. Questions and avenues for lucerne improvement. Czech Journal of Genetics and Plant Breeding. 2010;**46**:1-13

[69] Hipskind JD, Paiva NL. Constitutive accumulation of a resveratrol-glucoside in transgenic alfalfa increases resistance to *Phoma medicaginis*. Molecular Plant-Microbe Interactions. 2000;**13**(5):551-562

[70] Winicov I. Alfin1 transcription factor over expression enhances plant root growth under normal and saline conditions and improves salt tolerance in alfalfa. Planta. 2000;**210**: 416-422

[71] McKersie BD, Murnaghan J, Jones KS, Bowley SR. Iron-superoxide dismutase expression in transgenic alfalfa increases winter survival without a detectable increase in photosynthetic oxidative stress tolerance. Plant Physiology. 2000;**1224**:1427-1437

[72] Xie DY, Sharma SB, Paiva NL, Ferreira D, Dixon RA. Role of anthocyanidin reductase, encoded by *BANYULS* in plant flavonoid biosynthesis. Science. 2003;**299**:396-399

[73] Tabe LM, Wardley-Richardson T, Ceriotti A, Aryan A, McNabb W, Moore A, et al. A biotechnological approach to improving the nutritive value of alfalfa. Journal of Animal Science. 1995;**73**:2752-2759

[74] Carrillo C, Wigdorovitz A, Trono K, Dus Santos MJ, Castañón S, Sadir AM. Induction of a virus-specific antibody response to foot and mouth disease virus using the structural protein VP1 expressed in transgenic potato plants. Viral Immunology. 2001;**14**:49-57

[75] Pasquet RS. Cultivated cowpea (*Vigna unguiculata*): Genetic organization and domestication. In: Pickersgill B, Lock JM, editors. Advances in Legume Systematics 8: Legumes of Economic Importance. Kew, UK: Royal Botanic Gardens; 1996. pp. 101-108

[76] Tarawali SA, Singh BB, Peters M, Blade SF. Cowpea haulms as fodder. In: Singh BB, Mohan Raj DR, Dashiell KE, Jackai LEN, editors. Advances in Cowpea Research. Sayce, Devon, UK: Co-publication Intl Inst Tropical Agric (IITA) and Japan Intl Res Center Agric Sci (JIRCAS); 1997. pp. 313-325

[77] Singh AK, Varma N, Yadav SK, Mohanty A, Singh S, Singh S. Indian forage genetic resources: Perspectives and strategies. Progressive Agriculture. 2009;**9**(2):250-256

[78] Singh T, Malaviya DR, Kaushal P. Genetic analysis of some morphological traits in Egyptian clover (*Trifolium alexandrinum* L.) sustainable use of grassland resources for forage production, biodiversity and environmental protection. In: Roy AK, Kumar RV, Agrawal RK, Mahanta SK, Singh JB, Das MM, Dwivedi KK, Prabhu G, Shah NK, editors. Extended Abstracts 23rd International Grassland Congress. 2015. Paper ID: 1467

[79] Putiyevsky E, Katznelson J. Cytogenetic studies in *Trifolium* spp. related to berseem. I. Intra- and inter-specific hybrid seed formation. Theoretical and Applied Genetics. 1973;**43**:351-358

[80] Bhaskar RB, Malviya DR, Roy AK, Kaushal P. Evaluation of exotic Trifolium accessions for disease incidence and resistance. In: Abstr. Nat. Symp. on 'Grassland and Fodder Research in the New Millennium'; IGFRI, Jhansi. 2002. pp. 31-32

[81] Renfro BL, Sprague EW. Reaction of *Medicago* species to eight alfalfa pathogens. Agronomy Journal. 1959;**51**:481-483

[82] Borges OL, Stanford EH, Webster RK. Selection for resistance to *Stemphylium botryosum* in alfalfa. Crop Science. 1976;**16**:456-458

[83] Chandra A. Screening global *Medicago* germplasm for weevil (*Hypera postica* Gyll.) tolerance and estimation of genetic variability using molecular markers. Euphytica. 2009;**1693**:363-374

[84] Boscaiu M, Riera J, Estrelles E, Güemes J. Números cromosomáticos de plantas occidentales 751-776. Anales del Jardín Botánico de Madrid. 1997;**55**:430-431

[85] Alomar G, Mus M, Rosselló JA. Flora endèmica de les Balears. Palma de Mallorca: Consell Insular de Mallorca; 1997

[86] Zhou C, Han L, Pislariu C, Nakashima J, Fu C, Jiang Q, et al. From model to crop: Functional analysis of a STAY-GREEN gene in the model legume *Medicago truncatula* and effective use of the gene for alfalfa improvement. Plant Physiology. 2011;**157**:1483-1496

Chapter 9

Bana Grass Growing in Sub Saharan Africa

Zivanayi Matore

Additional information is available at the end of the chapter

http://dx.doi.org/10.5772/intechopen.82367

Abstract

The chapter provides a detailed summary on the morphology and physiology of Bana, climatic and soil adaptation, establishment, fertilization, weeding, basic management, and uses of the grass such as hay and silage making in sub-Saharan Africa. A detailed review of pest, disease, and weed control as well as grazing management of the crop is given at the end of the chapter. Bana grass is a very robust improved grass that has vast potential to improve animal production in the tropics. The grass was originally developed in South Africa as a cross between *Pennisetum purpureum* and *Pennisetum americum* and was widely used as livestock feed. However, the limited use of the grass in some parts of sub-Saharan Africa like Zimbabwe might be attributable to knowledge gap in production and overall importance of the grass. The high cost of commercial supplementations in ruminant animals particularly in sub-Saharan Africa justifies promotion of improved forage production. The author calls upon researchers, government structures responsible for agriculture, and development partners to promote establishment and utilization of the miracle grass in order to improve livestock production and livelihoods and reverse the terrible effects of climate change in sub-Saharan Africa.

Keywords: Bana grass, perennial grass, silage, pasture, dry matter, water-soluble carbohydrates

1. Introduction

Bana grass also known as the miracle grass is natural and invasive to Tropical Africa and the sub-Saharan region. The bunch grass is high yielding and deep rooted. Bana grass is a hybrid derived from the annual Babala (*Pennisetum americanum*) and the perennial Napier grass (*Pennisetum purpureum*) and was developed in South Africa as food for livestock.

The name Bana is derived from the acronym **ba** in Babala and **na** from Napier. However, some authors reiterate that the word Bana is from the word Ghana, the nation where it is abundantly grown. For better and maximum yields, the grass can be grown in warm climates that receive annual rainfall ranging from 750 to 2500 mm. Bana is susceptible to frost and therefore does not perform well in frost-prone regions. In Kenya and other African countries, Bana is the most common fodder crop where it is used to feed smallholder dairy cows and supplement beef animals during the dry winter [2]. With the high cost of commercial supplementation in dairy production, Bana grass provides a cost-effective way of improving milk production [4]. The popularity of Bana grass is associated with its drought-tolerant nature, its low nutrient requirement, and its ability to resist pests and diseases. In the past, Bana has been used chiefly for grazing; lately, however, it has been incorporated into a pest management approach and can also be used as a source of fuel and as wind break in homesteads or banana plantations. The grass has shown potential at attracting stem borer moths away from maize and hence is the "pull" crop [13]. This strategy is much more sustainable, serves more purposes, and is more affordable for farmers than insecticide use. Furthermore, Bana grass improves fertility of the soil especially when reinforced with legumes, and like many other robust grasses, Bana protects soil against erosion. It is also utilized for firebreaks, windbreaks, in paper pulp production, and most recently to produce bio-oil, biogas, and charcoal. The grass resembles sugarcane in appearance. It is characterized by pale green leaves and can grow up to 4 m. Bana grass is tolerant to drought and can be grown even on infertile lands. Bana is mainly propagated vegetatively, that is, through cuttings though some seed producing varieties have been reported but are not very common. Under good management, Bana yields up to 80 tons per hectare, but this can go down to 10 if fertilization and watering regime are poor. Frequent defoliation improves herbage quality but reduces herbage yield [12]. Knowledge gap in production and values of Bana limits Bana production in the subtropical regions. Literature publishing and dissemination will help alleviate this gap.

2. Bana botanical description and genetic variation

Kingdom: Plantae, **Order**: Poales: **Family**: Poaceae, **Genus**: *Pennisetum*, **Species**: *P. Purpureum*

2.1. Reproductive biology

Bana grass is mainly vegetative propagation, though sexual reproduction has been reported in some apomictic species of Bana grass. The apomictic species is an unpredictable seed producer, and in some environments, it seldom develops seeds, probably due to low pollen viability. Produced they are dispersed by wind but viability via this method is often very low.

2.2. Physiology and phenology

P. purpureum is a fast-growing perennial grass that flowers at different times of the year depending on climatic condition, for example, in Florida, the grass flowers between July and February. In Mexico and Central America, flowering occurs all year long with peaks from

December to May. In South Africa, this species flowers from January to June. As many other C4 grasses, *P. purpureum* is well adapted to environments with high daytime temperatures, intense sunlight, drought, and nitrogen and/or CO_2 limitation. Though the grass grows best during the rainy season, its deep root system allows it to survive long drought periods.

2.3. Morphology

Bana grass is a robust, rhizomatous, tufted perennial grass. The root system of Bana grass is vigorous and it develops from the nodes of its creeping stolons. The culms can reach a height of 4–7 mm and usually coarse and perennial. The grass forms dense thick clumps, up to 1 m across [6]. The leaves have a blue-green color and are flat, linear, and hairy at the base, with a length of 100–120 cm and a width range of 1–5 cm. The midrib for the Bana grass leaf can be easily seen with naked eyes, and the margins of its leaves are flimsily toothed. Bana produces no seeds at all; if ever seeds are produced, there will be very small seeds (3 million seeds/kg). Bana grass resembles sugarcane in appearance (*Saccharum officinarum*), the reason why some people call it sugarcane grass or elephant grass, but it is easily distinguished from sugarcane because of its leaves that are narrow leaves and the stems that are taller reaching up to a height of 4 m.

3. Uses of Bana

Bana grass is very important forage in the sub-Saharan Africa due to its high productivity. It can be used as silage or hay to feed dairy cows and supplement beef cattle and sometimes goats during the dry winter. Bana grass can be used under cut-and-carry systems ("zero grazing") or grazed, provided it can be kept at the lush vegetative stage: livestock tend to feed only the fresher leaves.

Bana grass is a multipurpose plant. The young leaves and shoots are edible by humans and can be cooked to make soups and stews. The yellow stem of Bana grass can be used to make fences and thatching houses (whole plant). The section gives a detailed outline of the use of Bana grass in subtropical Africa.

3.1. Silage

Bana grass is known for producing good-quality silage either alone or mixed with other crops. However, ensiling of Bana grass is complicated by the low concentration of water-soluble carbohydrates and its high cell wall content [6, 7]. Studies done in Zimbabwe, concluded that for good quality silage production. The optimal time for harvesting Bana grass should be when the concentration of water-soluble carbohydrates is at their pick point (6–7 weeks), to increase DM content and optimize herbage production without affecting nutritive value. When making Bana grass silage, the high moisture content of the grass when its nutritive value is highest is the main barrier for using Bana grass in silage making, because it results in undesirable fermentation with considerable nutrient losses. It is against this background that the grass is usually ensiled with materials that improve the quality of the silage and its nutritional value (protein or energy) such as cassava, velvet bean, and rice bran.

3.2. Hay and dry grass

When Bana grass is to be used as hay, though it is less common that it should be cut at an early stage of maturity as the stems become too coarse when the plant ages. In some parts of Africa, Bana grass is used for the production of dry grass pellets used as additional stock feed during the dry winter.

3.3. Fresh grass

The best harvesting regime for Bana grass is determined by weather conditions, inherent soil fertility, the levels of management, and physiological needs of animals. The 6- to 8-week harvesting interval can then be adopted. In Eastern African countries like Tanzania and Kenya and even Southern African countries like Zimbabwe, farmers are advised and recommended to harvest Bana grass for the first time when it attains a height of 1–1.2 m, usually 3–4 months after planting [2]. Thereafter, the grass should be harvested at intervals of 6–8 weeks, at the same height. Under excellent management, the grass can be harvested on a monthly basis in hot and wet environments or every 2 months in drier areas. If the grass is harvested at longer intervals, it produces higher DM yields, but the quality of the forage will be low as content of essential nutrients like protein and ash decreases with maturity of the grass. The digestibility of the grass will also go down; this is a direct result of the increase in the ratio stem-to leaf. When harvesting the grass as fresh forage under the cut and carry system, it is important to remember to leave behind the stubble with a height of 10–15 cm. This stubble will provide adequate energy reserves for subsequent regrowth.

Under cut-and-carry systems, Bana grass is often fed fresh to animals. To promote nonselection of leaves and stems and to increase voluntary feed intake, the grass can be chopped and fed to animals. Thus, for efficient forage utilization, the grass needs to be chopped and then sun wilted for several hours. This treatment of Bana decreases moisture, stimulates appetite, and facilitates rumination.

3.4. Pasture

When used for pasture, Bana grass should be heavily grazed so that most of the young leaves and shoots, which have the highest nutritive value, are available to ruminants. For the best utilization, the crop should be grazed at intervals of 6–9 weeks and at a height of about 90 cm. The crop can be top-dressed with nitrogen fertilizers after each grazing or cutting in high-rainfall areas.

3.5. Biological control agent of pests

Studies have shown that Bana grass in conjunction with other grasses and legumes like *Desmodium* spp. effectively controls notorious pests like maize stem borer moth. If planted alongside maize fields, the stem borer moth is pushed out of the field by *Desmodium and* will go and lay eggs on Bana grass [9]. When the larvae start boring Bana grass, the grass emits a sticky chemical that kills almost all larvae while the surviving ones are attacked by *Cotesia sesamiae*.

3.6. Weed and soil erosion control

Bana grass has been used successfully in erosion control, gully reclamation, and mulching of garden and field crops to fight against stem borers in maize crops. The grass is a good weed controller and, in Africa, it has been reported to be used as a trap plant in push-pull management strategies.

3.7. Climate and soil adaptation (ecology)

Bana grass is grown in both the tropical and subtropical regions. Rainfall amounts that range from 750 to 2500 mm per year and temperature range of 25–40°C are ideal for Bana grass production. However, it can be grown even in climates that receive less than 500 mm annual rainfall. Bana grass is prone to frost damage and it is not advised to grow it in frost-prone areas. Generally, the grass grows very well in regions that receive high rainfall, but its deep rooted system adapts it to drought risk areas that receive less rainfall.

Bana grass can be grown on a wide variety of soils that have good drainage, from the infertile sandy soils to the nutrient-rich loam soils. But it is best advised to grow Bana on unutilized soils with good drainage as it does not tolerate waterlogging and prolonged flooding. The grass can be grown on wet lands, flood plains, riverbanks, swamps, forest edges, disturbed sites, and waste ground. The grass grows well in soils with a pH range of 4.5–8.2, and there is no literature on tolerance of Bana to soils with high salinity.

Bana is found on vleis and river banks in the wild, explaining its need for good moisture, though it is not tolerant to prolonged flooding.

3.8. Establishment and yields

Bana grass is best planted in spring in temperate climate to give it the opportunity to grow before frost but throughout the year in sub-Saharan Africa but best planted during the rainy season to reduce irrigation requirements. Land preparation methods for planting Bana grass are the same as for maize. Identified land is plowed, disked, and harrowed before planting the grass [4]. Bana grass is best propagated vegetatively (stem cuttings with two to three nodes). These nodes are pushed into the soil at an angle of 45° with the bottom end down. Two nodes will then be buried into the soil. Bana is planted at an interrow spacing of 0.5–2 m and intrarow spacing of 0.3–1 m. However, when planting Bana grass for soil conservation purposes, close spacing is required. In the subtropical regions, since the annual average rainfall is usually low, it is advisable to increase spacing. The grass needs to be irrigated frequently soon after planting to establish the proper root system. The grass grows rapidly after planting and can grow up to 4 months soon after planting. Due to its fast growing nature, Bana grass has high yield potential that of course is influenced by prevailing climatic conditions. In the sub-Saharan Africa for Bana grass to fully realize its potential (20–80 t DM/ha/year), there is need to regularly irrigate the crop and heavily fertilize it. It is important to note that when management is poor, yields as low as 2–10 t DM/ha/year are realized. When space is limiting air layering or marcotting can be done to propagate Bana grass. Air layering is not different from the other forms of layering except that roots are induced on the part of the plant while it is still above the ground (air layering). When performing air layering for Bana grass, one node must be under the soil such that roots will develop from the nodes.

Bana grass can also be best planted by digging up a clump and then separating the segments with their roots. The separated segment can then be planted at an interrow spacing of 0.5–2 m and intrarow spacing of 0.3–1 m. The segments will also form clumps.

4. Harvesting of Bana grass (caring and pruning)

Harvesting intervals and technique for Bana depends with intended use. Bana grass can be harvested at intervals of 3–4 months depending on management and climate to pave way for renewed growth when feeding it livestock [1]. This can be done by allowing the animals to graze freely or the common cut-and-carry system. This cut-and-carry system is the best method of harvesting Bana and is normally referred to as the cutting back system. Under dry land or hard to irrigate conditions, Bana is cut back at the end of the rainy season to ground level, and the clumps will shoot up vigorously during the rainy season. It is important to note that under excellent management, harvesting intervals can go up to 6–8 weeks for 5 years [5]. When harvesting Bana under the cut-and-carry system, it is encouraged to leave a stem height of at least 10 cm from the ground to encourage vigorous regrowth. The grass can be fed green or as hay to livestock. When Bana is grown for making fuel, it can be cut every 9–12 months to increase the quantities of fuel obtained. It is important that even under poor management, Bana fares better than most grasses.

5. Intercropping Bana grass

Though Bana grass is mainly grown in pure stands, it can be intercropped with legumes and other shrubs such as *Leucaena leucocephala*, and intercropping Bana grass and leguminous improves the nutritional value of DM yield and soil fertility. Legumes such as Leucaena, Desmodium, Sesbania, and Mulberry can be intercropped with Bana. Some authors alluded *that Desmodium intortum* gives the DM content and nutrient value in intercropping regimes when compared to most legumes such as Sesbania, Leucaena, etc.

6. Nutritional attributes of Bana grass

Bana grass has rather low protein content (about 10% DM), but young grass can be very nutritive. For instance, studies in Venezuela revealed that at 30 days of regrowth, protein values ranged from 21% DM but reduced down to less than 4% DM when Bana was cut at 70 days. Bana grass is reported to be rich in fiber, but the fiber content is dependent on stage of maturity. The Neutral Detergent Fiber for Bana ranges from 55 to 75% depending on stage of maturity.

7. Potential constraints in use of Bana grass

7.1. Nitrate poisoning

When used as a sole component of the diet, Bana grass can cause nitrate poisoning in cattle. In Malaysia, it was reported that some cattle died from nitrate poisoning after they were fed with Bana grass. The levels of nitrate in Bana grass from the toxic area averaged 28.3 mg/g (up to 44 mg in some samples), while the levels of nitrates in Bana grass from nontoxic areas were 3.9 mg/g. When cattle manure was used to fertilize Bana in Brazil, two outbreaks of nitrate poisoning were reported. Some of the notable clinical signs in livestock suffering from nitrate poisoning after consuming Bana grass include uncoordinated gait, extreme salivation, anorexia, discharge from the nose, respiratory distress, grinding of teeth, depression, or abdominal contractions, cyanosis, and finally recumbency. The actual quantities of Bana grass that results in toxicity after ingestion are not yet known.

7.2. Mature leaves

When Bana leaves reach maturity stage, its leaves will be razor sharp and can occasionally hurt foraging animals. This is common in instances where cutting back rate will be low.

8. Animal production and Bana grass

8.1. Ruminants

Bana grass is one of the most vital fodder grasses for feeding ruminant in sub-Saharan Africa, mainly due to its high productivity. The grass can be grazed, fed as hay, cut fresh, and fed to animals or ensiled [13]. Voluntary intake of the grass is affected by variability among cultivars.

8.2. Palatability

Bana grass is very palatable during its early growth stages (young and leafy). However, as it matures, it becomes coarse and unpalatable. In light of this background, fresh elephant grass is often chopped to prevent animals from selecting the best parts.

8.3. Digestibility and intake

The high cell wall content of Bana grass reduces its protein and energy content. In situ digestibility trials revealed that there is a general decrease in crude protein content and an increase in the fiber content as the grass matures. Young Bana grass has a high nutritive value like most grasses [4]. Bana retains a high level of digestibility over a longer period because its cell wall content increases at a lower rate as it approaches maturity when compared to other grasses such as Kikuyu. Studies with steers in Brazil reported a decline in dry matter intake and organic matter digestibility with days of regrowth. OM digestibility varied from 75% at 33 days to 56% at 93 days. The authors then recommended use of Bana grass between 30 and 35 days of regrowth for effective organic matter digestibility and voluntary feed intake.

Due to a fill intake caused by water intake when Bana grass is very young, voluntary intake will be usually low. Some results suggest that, at restricted level of intake, maturity can result in an increase of metabolizable energy available in the gastrointestinal tract. Studies in Pakistan revealed that the use of N fertilizer increased the protein concentration of the Bana grass, but these fertilizers failed to reverse the adverse effects of maturity on nutrient digestibility in buffaloes. Bana grass can be chopped and roller-milled to increase voluntary feed intake of the grass and overall value of the grass and consequently a reduction in feed costs. This is made possible by the increase in cell wall surface area available for digestion by microbes in the rumen.

9. Some livestock classes and feeding of Bana grass

9.1. Dairy cattle

In most smallholder dairies in sub-Saharan Africa, Bana grass is a popular forage and is used to feed dairy cattle under the cut-and-carry system [2]. The grass is cut at a height of 55–60 cm, that is, 2–3 months after planting and fed to dairy cows [11]. To reduce feed wastage, the grass needs to be chopped into smaller pieces. A mature dairy cow will consume about 10 kg of Bana a day under intensive management. Effective utilization of freshly harvested Bana is affected by its low dry matter content and high fiber content. When supplemented with Leucaena, Bana grass will sustain milk yields of 7–8 l/day. For adequate performance of dairy cows, the grass should be supplemented with leguminous forages.

9.2. Sheep and goats

Sheep and goats raised for either mutton/chevron production can be sustained by Bana [8]. However, for lactating goats and sheep, there is need to supplement this class of stock with leguminous forages. Legumes supply protein to animals, which is an important component in milk synthesis.

9.3. Rabbits

In tropical countries such as Vietnam, Nigeria, and Mozambique, Bana grass is commonly used as green forage for feeding rabbits. When fresh, Bana grass is moderately palatable to rabbits and is associated with high productivity. Palatability of Bana grass to rabbits is also affected by factors such as season, with high levels of palatability and crude protein during the rainy season, while low palatability during the dry season. Bana grass needs to be fed together with other legume forages such as Lucerne for effective growth.

10. Diseases and pests

Bana grass just like other grasses and crops can be attacked by various disease-causing agents, but it has shown a greater degree of resistance compared to other forages [10]. In parts of sub-Saharan Africa like Central Kenya, reports of the grass being attacked by head smut caused by *Ustilago camerumensis* were received [3]. However, some varieties of Bana were found to be resistant to the disease. *Helminthosporium* spp. has also been reported in Bana, but a lot of work is being done to look for varieties that are tolerant to this disease-causing agent. During the rainy season, now mold fungal disease is common to most species of Bana grass with the exception of Clone3. Nevertheless, herbage production is not threatened by snow mold fungal disease.

11. Source of planting material

In most sub-Saharan African countries, Bana grass planting material can be obtained from various places and sources such as research institutions, farmers who are into production, ministry of agriculture, agriculture colleges, and universities.

12. Limitations in Bana grass production

Despite its robust growth and the big production potential, Bana has its own limitations. Bana matures rapidly and becomes stemmy, making it highly unpalatable if growth is not controlled [2]. Bana is also propagated vegetatively and rarely from seeds, since most varieties produce seeds that are not viable. In order for Bana to realize its maximum yield, it needs to be established on soils with high inherent fertility or on heavily manured soils.

13. Conclusion

Bana grass has always had a valuable role in the world of agriculture over the years. The potential of the grass for improving agriculture and bringing stability to the ecosystem is enormous in sub-Saharan Africa. Nevertheless, its future use in agriculture is limited by the knowledge gap and poor adoption in general. The current impacts of climate change together with the high cost involved in feeding animals justify the need to improve establishment and utilization of the miracle grass. Bana grass is relatively easy to propagate and can reach up to a height of 3 m in 3–4 months and has high leaf to stem ration compared to most improved grasses. The grass can also be grown on hard-to-irrigate areas but of course on soils with good irrigation, making it suitable for the sub-Saharan climate. The grass is also native to sub-Saharan Africa. The author calls upon researchers, government structures responsible for agriculture, and development partners to promote establishment and utilization of the miracle grass in order to improve livestock production and livelihoods and reverse the terrible effects of climate change in sub-Saharan Africa.

It is preferable to include a Conclusion(s) section, which will summarize the content of the book chapter.

Acknowledgements

I thank you lord for thy hand of protection, for providing resources and time as well as for the divine guidance throughout the whole process of writing this chapter. Without your grace and love, the dream was never going to be a reality.

To my workmate Mr. I Dzuke: I thank you for the moral and resource support you gave during the time of writing the chapter.

To my family (Violet my better half, Tinotenda the blue eyed boy, Sindisiwe my beloved daughter): I thank you for your words of encouragement that saw me pushing through. I sincerely thank you for your prayers that contributed to the successful writing of this book chapter.

Conflict of interest

Matore Z, the author of this book chapter, states that there is no conflict of interest in this book chapter.

Author details

Zivanayi Matore

Address all correspondence to: zetmatore@gmail.com

Crops and Livestock Department, Gweru, Zimbabwe

References

[1] Rengsirikul K, Ishii Y, Kangvansaichol K, Pripanapong P, Sripichitt P, Punsuvon V, et al. Effects of inter-cutting interval on biomass yield, growth components and chemical composition of napiergrass (*Pennisetum purpureum* Schumach) cultivars as bioenergy crops in Thailand. Grassland Science. 2011;**57**(3):135-141

[2] Orodho AB. The Role and Importance of Napier Grass in the Smallholder Dairy Industry in Kenya. Rome: Food and Agriculture Organization. Retrieved August 24 (2006): 2011apier_kenya.htm. [Accessed: May 20, 2016]

[3] Asudi GO, Van den Berg J, Midega CAO, Pickett JA, Khan ZR. The significance of Napier grass stunt phytoplasma and its transmission to cereals and sugarcane. Journal of Phytopathology. 2016;**164**(6):378-385

[4] Turano B, Tiwari UP, Jha R. Growth and nutritional evaluation of napier grass hybrids as forage for ruminants. Tropical Grasslands-Forrajes Tropicales. 2016;**4**(3):168-178

[5] Tessema ZK, Mihret J, Solomon M. Effect of defoliation frequency and cutting height on growth, dry-matter yield and nutritive value of Napier grass (*Pennisetum purpureum* (L.) Schumach). Grass and Forage Science. 2010;**65**(4):421-430

[6] Budiman B, Soetrisno RD, Budhi SPS, Indrianto A. Morphological characteristics, productivity and quality of three napier grass (*Pennisetum purpureum* Schum) cultivars harvested at different age. Journal of the Indonesian Tropical Animal Agriculture. 2012;**37**(4):294-301

[7] Mendieta-Araica B, Spörndly E, Reyes-Sánchez N, Norell L, Spörndly R. Silage quality when *Moringa oleifera* is ensiled in mixtures with elephant grass, sugar cane and molasses. Grass and Forage Science. 2009;**64**(4):364-373

[8] Kozloski GV et al. Intake and digestion by lambs of dwarf elephant grass (*Pennisetum purpureum* Schum. cv. Mott) hay or hay supplemented with urea and different levels of cracked corn grain. Animal Feed Science and Technology. 2006;**125**(1-2):111-122

[9] Bowden J. Stem-borer ecology and strategy for control. Annals of Applied Biology. 1976;**84**(1):107-111

[10] Asudi GO, van den Berg J, Midega CA, Pittchar J, Pickett JA, Khan ZR. Napier grass stunt disease in East Africa: Farmers' perspectives on disease management. Crop Protection. 2015;**71**:116-124

[11] Rusdy M. Elephant grass as forage for ruminant animals. Livestock Research for Rural Development. 2016;**28**(4)

[12] Palacios-Díaz MP, Mendoza-Grimón V, Fernández-Vera JR, Hernández-Moreno JM. Effects of defoliation and nitrogen uptake on forage nutritive values of *Pennisetum* sp. Journal of Animal and Plant Sciences. 2013;**23**(2):566-574

[13] Negawo AT, Teshome A, Kumar A, Hanson J, Jones CS. Opportunities for Napier grass (*Pennisetum purpureum*) improvement using molecular genetics. Agronomy. 2017;**7**(2):28